AF776229

The History of Grinding

Alban J. Lynch
Chester A. Rowland

Published by the
Society for Mining, Metallurgy, and Exploration, Inc.
8307 Shaffer Parkway
Littleton, Colorado

Dedication

We dedicate this book to our wives, Barbara Lynch, Lucy Rowland, and the late Anne Rowland, who helped to make this book possible. During our professional careers, we were away from home a great deal–collecting operating data for crushing and grinding circuits, conducting crushing and grinding research, supplying consulting services, and attending technical meetings and conferences–while Barbara and Anne raised, respectively, seven and three children. We appreciate the patience and understanding of Barbara, Anne, and Lucy throughout our careers and during the 5 years we spent writing this book and show our gratitude and admiration by dedicating this book to these three remarkable women.

Society for Mining, Metallurgy, and Exploration, Inc. (SME)
8307 Shaffer Parkway
Littleton, Colorado, USA 80127
(303) 973-9550 / (800) 763-3132
www.smenet.org

SME advances the worldwide mining and minerals community through information exchange and professional development. SME is the world's largest association of mining and minerals professionals.

Copyright © 2005 Society for Mining, Metallurgy, and Exploration, Inc.

All Rights Reserved. Printed in the United States of America.

Information contained in this work has been obtained by SME, Inc., from sources believed to be reliable. However, neither SME nor its authors guarantee the accuracy or completeness of any information published herein, and neither SME nor its authors shall be responsible for any errors, omissions, or damages arising out of use of this information. This work is published with the understanding that SME and its authors are supplying information but are not attempting to render engineering or other professional services. If such services are required, the assistance of an appropriate professional should be sought.

No part of this publication may be reproduced, stored in a retrieval system, or transmitted in any form or by any means, electronic, mechanical, photocopying, recording, or otherwise, without the prior written permission of the publisher. Any statement or views presented here are those of the author and are not necessarily those of SME. The mention of trade names for commercial products does not imply the approval or endorsement of SME.

ISBN 0-87335-238-6

Library of Congress Cataloging-in-Publication Data.

Lynch, Alban J., 1930-
The history of grinding / Alban J. Lynch, Chester A. Rowland.
p. cm.
Includes bibliographical references and index.
ISBN 0-87335-238-6
1. Grinding and polishing--History. I. Rowland, Chester A., 1924- II. Title.

TJ1280.L96 2005
671.3'5--dc22

2005040169

Contents

Preface

During our professional careers, we placed our principal emphasis on understanding the role and use of energy in size reduction, particularly in the areas of crushing and grinding minerals, the use of simulation for crushing and grinding circuits and machinery, and the design and use of the ever-larger size-reduction machinery.

In 1998 while discussing writing our memoirs, we realized that the task would likely focus on the spectacular changes that had occurred in crushing and grinding in the mineral industry during the second half of the 20th century. As we began to write, we were able to understand why mineral processing engineers in 1948 had found it difficult to envision the size and scope of mineral size-reduction processes and equipment in 1998. At first, we thought that writing the memoir would be a simple task; then we started to put the 50-year period from 1948 to 1998 into the context of a process that human beings had used for hundreds of thousands of years and that had been changing and evolving for 4,000 years. Our simple memoir turned into a fascinating study of how size-reduction technology developed over not 50, but thousands of years.

A memoir that we expected to write in a few months took more than 5 years to complete and resulted in this book. Here, we have attempted to outline, without becoming immersed in technical detail, how size reduction evolved and have illustrated developments with photos and drawings from many publications, some more than 100 years old. We are well aware that this book is incomplete, with some of the omissions including, for example, grinding wood to make paper; grinding and sanding the surfaces of metals and other materials that require cleaning, smoothing, and polishing; and coal cutting.

ACKNOWLEDGMENTS

Our colleagues–some of whom we have known for many years and some we have come to know recently through e-mail–have responded most generously to our requests for information and comments. We are grateful for all the assistance we received. We would like to extend our particular thanks to those colleagues who wrote or contributed to sections in the book. In particular, we acknowledge the contributions of Doug Schlepp and Bud Green from Krebs Engineers (Tucson, Arizona), who wrote most of the section on hydrocyclones. Thanks also to Klaus Schonert from the Technical University of Clausthal, Germany, for the description of the work that led to the development of high-pressure grinding rolls. Philip Lloyd, a consultant in South Africa, gave us the history of the development of the planetary mill at the Chamber of Mines in South Africa. The Dorr Oliver Company sent us information on J.V.N. Dorr; and Stuart Jones of Metso Minerals, Grinding Section, supplied the historical information on Harlow Hardinge.

A.J.L. wishes to thank the many colleagues with whom he discussed aspects of blasting, crushing, and grinding in Wordsmith's coffee shop at the University of Queensland and Gossip's coffee shop at Toowong village, in addition to Vynette Halliday for her assistance with diagrams.

With the figures included in the book, we have acknowledged the permissions we received to publish figures that we extracted from published information. Where we were unable to contact the original publisher to obtain permission to publish, when possible we acknowledge the source for the figure.

We are indebted to our editor, René Howard. She was a tremendous help in helping us organize the book, which started out as a series of disconnected technical reports. It was René's guidance, encouragement, and patience that made the completion of this book a fruitful experience.

CHAPTER 1

Size Reduction from the Stone Age to the Space Age

CHIMPANZEES AND NUT CRACKING

The history we give in this book begins at the start of agriculture, about 8000 BC, and it refers only to the activities of Homo sapiens. Size reduction, however, may actually have been pioneered long before that. For example, chimpanzees in some regions of West Africa have a culture of nut cracking that may extend back 2.6 million years:

> *In the Tai forest (Ivory Coast) mother chimpanzees teach their infants the tasty art of nut cracking which takes about seven years to master because about 1,000 kg of skillfully applied force are required to split the nut without pulverizing it. ...The stone hammers—rocks—acquired by the chimps are carried to specific nut-cracking sites where the chimps pound the golf-ball-size nuts of the* Panda oleosa *tree to expose the three seeds within. (Trivedi 2002)*

> *Powerful hits are required to initially crack the thick outer shell of the* Panda *nut; then a series of hits are precisely graded to crack the inner shells without shattering the almonds. The nut must be repositioned at least three times during this process. (Calvin 1991)*

HOMINIDS AND SIZE REDUCTION

Size reduction—the oldest engineering process—had its beginnings in prehistoric times, when early humans pounded grains and nuts with stones to free the edible inner parts from the hard protective shells. The mortar and pestle, a combination of a bowl-shaped container (the mortar) and a rod with a flat-surfaced hammer (the pestle), was the earliest tool used for this task, and, in principle, has remained unchanged throughout the ages. It is still used today in chemical laboratories and pharmacies.

> *The remains of seven types of 780,000-year-old nuts have been found at the Gesher Benot site in Israel's Hula Valley. The nuts and the stone tools found with them are the first evidence that various types of nuts formed a major part of man's diet 780,000 years ago and that hominids had developed an assortment of tools to crack open nuts during the Early-Middle Pleistocene Period. ...Some of the stones are the size of hammers while larger stones, some weighing as much as 30 kg could be used as anvils. (Goren-Inbar et al. 2002)*

> *From Palaeolithic times onwards man has used stones to disintegrate food-stuffs like acorns, nuts and wild grasses by using a pear- or pestle-shaped stone to beat and rub the material laying on a flat stone sometimes showing a natural or an artificial depression. (Forbes 1955)*

> *During historic ages the globe had passed through four of the six thousand years of its recorded technology before any device other than the hand-stone was known for reducing the size of grains and breaking nuts. (Bennett and Elton 1898–1904)*

The hand stone is still essential for producing food in many communities. Its importance was never better recognized than by the choice of a hand grinding stone as the centerpiece of the Women's Memorial in Pretoria, South Africa, to commemorate the centuries of arduous work by South African women (W. Cruise, personal communication; see Figure 1.1).

Stone grinding of ocher for pigments is also ancient. More than 40,000 years ago, Australian aborigines extracted high-grade ochers from iron oxide deposits and ground them with stones to a powder for use in cosmetics, body and artifact decorations, and cave paintings.

According to Fischer (1944), "In gradually changing form the world's oldest machine (the grinding mill) has served men in every country for more than 100 centuries."

THE EARLIEST TECHNOLOGIES

About 4,000 years ago, with the mortar and pestle proving unsuitable for the finer grinding needed for new commodities, the saddlestone mill was invented. Saddlestones use a horizontal fixed stone, over which the moving stone travels in a back-and-forth motion. This mill, which again relied on human muscles as the source of energy, was used to finely grind various raw materials for more than 1,500 years. But working with this mill was exhausting, and grinding became a feared form of labor, hated so deeply that it had to be carried out by slaves, prisoners, and the most menial of servants.

> *You are now a slave. Turn the millstone, grind the flour, off with your veils, strip off your fine clothes.* (Good News Bible *1976, Isaiah 47:2)*

Millstones were essential to daily domestic life in biblical times, and the meaning of any reference to them in a story or parable was understood instantly:

> *When you lend a man something, you are not to take as security his millstones used for grinding his corn. This would take away the family's means of preparing food to keep alive.* (Good News Bible *1976, Deuteronomy 24:6)*

> *Then a mighty angel picked up a boulder the size of a large millstone and threw it into the sea saying "this is how the great city of Babylon will be violently thrown down and will never be seen again" (Revelation 18:21); ...but a woman threw a millstone down on his head and fractured his skull... (Judges 9:53); [and] ...his stony heart is without fear, as unyielding and hard as a millstone (Job 41:24).* (Good News Bible *1976)*

References to millstones continued in stories and legends until roller mills replaced millstones and the stones lost their mystique. For example, in "The Juniper Tree" in Grimm's Fairy Tales, the cruel stepmother received her just desserts for ill-treating her stepchildren when "the bird let fall the millstone upon her head and crushed her to pieces" (Grimm Brothers 1993). The place of millstones in folklore will forever be assured:

> *St. Piran, the patron saint of Cornish tinners, according to folklore, sailed to Cornwall from Ireland on a millstone. It seems he offended one of the Irish kings who had him shackled to a millstone and thrown into the sea. Those watching were amazed to see him rise to the surface of the water sitting astride the millstone. [The story continues that St. Piran introduced tin extraction to Cornwall when he saw the white metal flowing from fissures in rocks that he was heating to cook his dinner.] (Jolliffe 2001)*

FIGURE 1.1 The grinding stone that is the centerpiece of the Women's Memorial in Pretoria, South Africa (reprinted by permission from Wilma Cruise)

Millstones were the principal tools used for size reduction until about 2500 BC, when the quern was invented. The quern's rotary motion was much easier to maintain than the back-and-forth motion of saddlestones. Although it could not match the saddlestone in making very fine particles, in time, the quern became widely used. Larger querns were developed that could be driven by animals, and, after the water wheel was invented, it was used to drive these larger querns. Although waterpower was the first renewable source of energy to be harnessed, 500 years would elapse before water wheels replaced muscles as the main source of energy for producing flour. Slaves, after all, were cheaper than machines and could solve processing problems as they arose.

The grinding of minerals lagged behind that of grain, because hard, abrasive pebbles tended to wear and break the expensive stones used in querns. For thousands of years, miners used chisels, hammers, picks, or levers as tools for freeing metallic and nonmetallic minerals embedded in rock formations. These tools were pounded into rocks to break them free from larger rock formations; the broken rock was pounded with smaller rocks or hammers; and the small pebbles were ground with the same hand tools used to grind grain. Fire was used to heat rock formations in the hope that they would crack as they cooled, although this was a costly use of fuel (particularly trees).

Although fireworks were the origin of explosives for breaking boulders and pebbles from rock masses, how and when fireworks were discovered is not known. They may have originated in China during the first millennium AD from peasants building fires on saltpeter rocks and finding that large volumes of gas were generated, carrying glowing particles into the sky. By 1250, saltpeter was being used for military purposes (Atlas Powder Company 1987), but it was 1627 before black powder, which is a saltpeter-charcoal-sulfur mixture, was used for breaking rocks in mines. Many mine owners hesitated to use it because of premature explosions, but the more adventurous owners used it successfully, leading to its widespread adoption. Black powder was the main explosive for mining until Alfred Nobel made dynamite around 1860. The dynamite era lasted 100 years until dynamite was replaced by the safer and more flexible ammonium nitrate–fuel oil (ANFO) mixture.

Drilling holes in rocks so that explosives could be placed where they would have the most effect proved to be one of the most intractable of all problems associated with breakage. Muscles provided the only power that could be used for many years before mechanical rock drills were invented. Starting in about 1860, drills began to be developed in parallel with explosives.

The size reduction of minerals changed about 1500 AD, when high demand for metals led to the development of water-driven stamp mills for grinding large volumes of pebbles into small particles. In stamp mills, the pebbles were shattered by impact from hammers, with iron protecting the wearing surfaces. Even though these mills were mechanically complex, they were unchallenged as fine-grinding machines for 400 years, particularly because they could grind pebbles quickly.

INCREASING POPULATION AND NEW COMMODITIES

As the world's population steadily increased (see Table 1.1), so did the demand for converting raw materials into valuable commodities such as medicines, minerals, and pigments. By the start of the 20th century, these and other commodities were being produced in volumes of tons (see Table 1.2).

Along with the spectacular increases in population and consumption came an ever-increasing demand for size reduction. Consider copper as an example. In 1900, 1 ton of copper could be produced from 50 tons of ore because the ore averaged 2% copper, but, by 2000, more than 100 tons of copper ore were required because the average copper content was less than 1%. As a consequence, the amount of grinding required has more than doubled, because more grinding is required to liberate the copper minerals, which are smaller in the lower-grade ores than in the higher-grade ores that are no longer available.

Throughout the history of size reduction, innovation has been driven mainly by technological advances associated with energy and with new materials. The only way to meet the increasing demands for flour and metals for thousands of years was to increase the number of small machines being used and the number of people working them. Only when larger amounts of energy became available from renewable resources such as water and wind could higher capacity, more-efficient machines be built. The effect was dramatic—more energy per machine meant more flour and less effort by human muscles. Both effects dramatically improved human quality of life.

Higher power and higher productivity per machine also meant higher stresses on the grinding surfaces. Hard, abrasion-resistant volcanic stones, which were suitable for soft grains and even for hard, abrasive ores at low feed rates, were used as the surfaces exposed to wear in size-reduction machinery until better wear-resistant materials were developed. When more energy became available and new machines and higher ore feed rates were introduced, better materials were developed to withstand the wear.

LANDMARK INNOVATIONS IN GRINDING

The 16th, 19th, and 20th centuries brought seminal changes in size-reduction technology. In each case, abundant new wealth encouraged the consumption of materials, investments in new plants, and the development of new processes and machines. This, in turn, increased the demand for size-reduction machinery with greater productivity and greater energy efficiency.

In the 16th century, the Renaissance in Europe and the silver and gold flowing to Europe from New Spain in Central America created the wealth. Gold became the standard for establishing value when trading goods. To this day, possessing gold—either as the metal itself or as fashioned into jewelry, statues, tableware, or household decorations—is a measure of personal wealth and status.

During the 16th century, waterpower replaced muscle power for grinding, and black powder was used as an explosive for removing boulders from fractured rock masses.

The development of the steam engine led to the advent of the railroad and steam-powered ships, which replaced horses and wind-catching sails as the driving energy for

TABLE 1.1 World population during the last 1,000 years and future projections

Year	1000	1500	1800	1910	1930	1950	1970	1990	2000	2010	2030	2050
Population, billions	0.30	0.50	0.98	1.75	2.07	2.52	3.70	5.30	6.23	6.85	8.25	9.35

Source: Infoplease.com 2003.

TABLE 1.2 Consumption of raw material–based products, 1900–2000

Product	1900	1930	1960	1990	2000
Gold, tons	448	746	1,049	2,133	2,576
Copper, millions of tons	0.50	1.06	4.10	9.00	14.56
Coal, millions of tons	780	1,100	2,000	3,500	5,300
Cement, millions of tons	NA*	200	500	1,000	1,700
Wheat, millions of tons	NA	145	220	592	594
Phosphate, U.S., millions of tons	1.52	4.02	17.80	46.3	38.6

* NA = not available

long-distance transportation by land and sea of people, materials, and goods. The use of steam power heralded the start of the 19th century's Industrial Revolution, which created additional wealth and dramatically expanded the production of consumer and capital goods and rapidly increased the demand for minerals. During this period, many machines were invented for size reduction.

Dynamite eventually replaced black powder for breaking rocks to yield raw materials, and a new generation of mining and milling machines was invented. The development of steam power—a reliable, concentrated form of energy—started a major revolution in the distribution of materials, food, fuel, and finished products, as well as in the transportation of people. Rail transportation, first with steam locomotives and then with diesel-powered locomotives, became commonplace.

By the start of the 20th century, demand for iron, steel, copper, lead, and other metals was strong. As the century unfolded, progress in electrical, mechanical, and materials engineering made bigger and better size-reduction machines possible, and the inexorable increase in population and demand for raw materials made them necessary. At that time, the newest sources of energy—electricity and oil—were generating great wealth. In addition, the population was increasing at an astonishing rate. The demand for products that resulted from greater wealth and more people was propelling the demand for size-reduction machinery with greater productivity and energy efficiency. For example, the development of automobiles and trucks driven by internal combustion engines, and later, the dramatic growth and flexibility of the airplane, meant that more materials needed to be reduced to ever smaller sizes at ever higher efficiencies. During this era, immense size-reduction machines were constructed to meet industry needs. Principally, these industries produced flour, construction stones and aggregates, metallic and non-metallic minerals, portland cement, and chemicals and fine powders.

The portland-cement process, the cyanide and flotation processes for minerals, and the gradual reduction process for cereal grains—all of which required efficient, high-capacity fine-grinding circuits—had been developed by the beginning of the 20th century. These processes set the stage for advances in size-reduction technology that would be made possible by the electrical age.

THE LINK BETWEEN ENERGY AND SIZE REDUCTION

Because size reduction depends totally on energy used—more energy means more breakage—the link between the two merits special mention (see Table 1.3). Progress in size-reduction machines has always depended on available sources of energy. First, it was necessary to supplement human muscle power as an energy source and then to replace it entirely. Tools that used animals—horses and oxen—to drive grinding devices were developed. Next, water wheels and windmills were invented, followed by motors powered by steam. Finally steam-driven turbines—now a source of energy around the world—were developed to generate electricity. This largely eliminated the use of human energy to drive size-reduction tools and machinery (except in some developing countries, where grain is still ground by hand).

Today, grinding mills use about 2% of the electricity generated in the world, and dry-grinding processes consume about 75% of this energy. (Note that some of the major size-reduction processes that consume a great deal of energy, such as the grinding of trees to produce wood pulp cellulose, are beyond the scope of this book.)

Size-reduction machinery is generally described as mechanical breakage mechanisms capable of applying available power to break down material from a specified feed size to a specified product size. The following key points summarize the types of energy used for size reduction as they developed over the years:

- The fuel for muscle power is the food eaten by humans and animals. Muscle-powered machines work at about 0.5 kW (0.66 hp).
- The fuel for waterpower is from the flow of water that drives the water wheel. Water-powered machines work at about 5 kW (6.6 hp).
- The fuel for steam power is the wood, coal, or petroleum that, when combusted, provides the heat to convert water to steam. Steam-powered machines work at 50 kW (66 hp).
- Electrical energy is produced in electrical generators driven by water, steam, or internal combustion engines. Effectively, there is now no limit to the power that can be delivered by electric motors.

THE PATH OF TECHNOLOGY

Over the centuries, great technical achievements led to the machines of today, which can grind solid particles at the rate of tens of thousands of tons per day. The evolution of machines from devices driven by one person to immense machines with capacities of many hundred tons per hour driven by motors larger than 20,000 kW (27,000 hp) has followed the path of technology in general, encapsulated by Henry Hodges (1970) in the final chapter of his book entitled *Technology in the Ancient World*:

> *Nowhere was the rate of technological advance a steady, even, upward climb. Always one seems to be confronted with sudden bursts of technological innovation followed by long periods of virtual stagnation.*
>
> and
>
> *If the history of technology in the ancient world is to teach us anything at all it is to tell us that no technology can stand long on its own in glorious isolation. For a technology to thrive and develop required the presence of other often quite unrelated technologies alongside it from which ideas could be lent or borrowed. Such was the case, anyway, in the ancient world. One is left wondering whether the world of today is so very different.*

TABLE 1.3 Energy consumed in grinding some widely used products

Material	Energy Used, kW-hr/t
Hard coal in power plants	7.6
Cement	67.5
Iron ore	40
Nonferrous ore	~30
Phosphate ore	~40
Inorganic pigment	880
Organic pigment	220
Wood pulp cellulose	1500
Cereals	26.5
Sugar, cacao, dried milk	45
Burnt lime	15

Source: Rumpf 1977.

THE CONTRIBUTION OF TECHNICAL AND NONTECHNICAL FACTORS

Key Technical Factors

- Processes were developed late in the 19th and early in the 20th centuries for producing cement and metals. These techniques—the McArthur–Forrest process (invented by J.S. McArthur and R. and W. Forrest) for leaching fine-grained gold ores, the flotation process for concentrating sulfides in fine-grained metal ores, and the rotary kiln for making cement—required finer particles than the mills at the time could produce. The ball mills built for these processes eventually grew into the huge machines of the 21st century.
- Electrical and mechanical engineering improvements allowed compact, high-power motors to be built and produce power to be transmitted efficiently into mill drives.
- High-quality steels that were resistant to wear, impact, and fatigue were invented; these reduced equipment breakdowns and increased production rates.
- Scientific advances led to better equipment and techniques for design and control.

Key Nontechnical Factors

- When the economic regime that controlled taxation and the availability of capital and skilled labor was favorable, owners of plants in which grinding mills were used invested in new equipment, including new sources of energy and materials.
- A vibrant mood of intellectual inquiry, which had been evident during the prosperous years of the Greek and Roman empires, the Renaissance, and the Science Age in the 19th century, emerged again during the second half of the 20th century, leading to much progress in size-reduction technology in Europe and the United States.

Karl Marx had no doubt about the importance of the grinding mill to the history of mechanical technology itself: "...the history of the development of machines may be retraced by the history of the flour mill" (Kuprits 1967).

As we review the history of size reduction, we should keep in mind that both technical and nontechnical factors came into play in the technology's evolution, particularly in the 20th century (see sidebar). In the rest of this chapter we review how sources of energy, quality of materials, and investments in technology changed the practice of size reduction (see Table 1.4 and Figure 1.2).

TABLE 1.4 The progress of size-reduction technology*

Stages in Technology Development	Factors Interacting at Each Stage
Human muscles	Available materials
Men or animals working together	Accumulated skills and experience
Waterpower	Economic and social conditions
Steam power	Religious and ethical tenets
Electricity	Philosophical doctrines
Energy Available per Machine	**Power Delivered by Muscles, ft-lb/sec**
Stone Age—Human muscles: 0.1 kW†/person	Draft horse: 432
600 BC—Men or animals working together: 0.5 kW	Ox: 288
100 BC—Water wheel: 5 kW‡	Mule: 216
1800—Steam engine: up to 50 kW§	Donkey: 108
1900—Electricity: 30,000+ kW (no limit)	Human pumping: 33
	Human winching: 45

Source: Adapted from Forbes 1955.

* We do not include explosive energy in the table because its growth has been separate from the growth of mechanical size reduction; however, explosives have been crucially important in size reduction, and the same interacting factors apply.

† 1 hp (0.746 kW) = 500 ft-lb/sec.

‡ Large water wheels generated more than 50 kW, but 5 kW seems a reasonable estimate for power used by many grinding machines operated by waterpower.

§ 50 kW is a reasonable estimate for steam engines used to drive grinding machines.

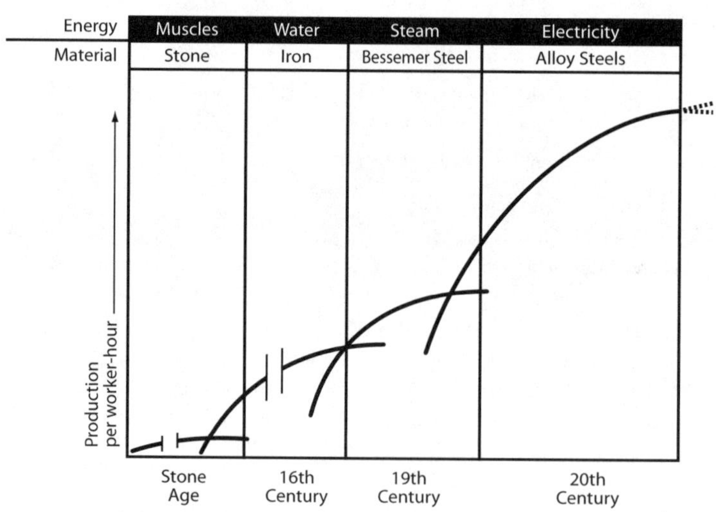

FIGURE 1.2 Effects of energy sources and materials on size-reduction processes

Table 1.5 gives a brief chronology of size-reduction technology, and Table 1.6 presents a more detailed chronology. (Italics denote an event that contributed to substantial progress in size-reduction technology.)

TABLE 1.5 Brief chronology of the development of energy used for size-reduction technologies

Time Frame	Technology	Effects
	Muscle Power	
Stone Age	Mortars and pestles	Pounding action; good for coarse grinding
2000 BC	Saddlestones	Reciprocating action; good for fine grinding
500 BC	Querns	Rotary action; good for grinding soft grains, less satisfactory for grinding hard ores
	Waterpower	
1500 AD	Stamp mills driven by water wheels	Pounding action; good coarse-grinding machine
	Steam Power	
1850–1900	Jaw crushers, ball mills, air classifiers	High-capacity machines for ores and cement
	Roller mills for grains	Precise machines to grind grains incrementally
	Electricity	
1900–1920	Vertical roller mills, rake classifiers, autogenous pebble mills	Closed grinding circuits for ores, coal, and cement; stones used as grinding media
1930–1955	Primary autogenous mills	Rocks or ore used as grinding media
1960–1975	High-diameter, short-length semiautogenous mills	Eliminated crushers in iron and copper ore circuits
1985–1999	Large tumbling mills, high-speed stirred mills	Very high-capacity and very fine-grinding machines

TABLE 1.6 Full chronology of size-reduction technology

Source of Power	Time Frame*	Machines and Processes
		Muscle Power
People	Stone Age	Mortars and pestles; hand stones; rock masses broken by fire and stone hammers
	2500 BC	Hand-operated bellows for compressed air
	2000 BC	Saddlestone mills
	1500 BC	Foot-operated bellows for compressed air
	500 BC	Rotary querns
Cattle	200 BC	Rotary querns
		Water Wheels and Windmills
Horizontal wheel	100 BC	Water-driven rotary querns, initially in east Mediterranean
Vertical wheel	20 AD	Water-driven rotary querns, initially in Roman Empire
Windmill	700	Wind-driven rotary querns, initially in Persia, Afghanistan
	200	Edge roller mills in China for fine grinding
Vertical water wheel	400	Waterpower replaced muscle power for grain milling in Roman Empire
	800	Dry stamp mills (manual power)
	1320	Black powder used for military purposes
	1512	Water-driven wet stamp mills
	1546	*De Re Metallica:* Agricola's textbook on mining engineering

TABLE 1.6 Full chronology of size-reduction technology (continued)

Source of Power	Time Frame*	Machines and Processes
	1558	Corrugated roller mill for grains—not used for 300 years
	1587	Amalgamation process for silver extraction
	1589	Arrastra for grinding and mixing in amalgamation process
	1627	Black powder used for mining purposes
	1776	Mechanical air compressor
	1785	Oliver Evans' automatic flour mill
		Steam
Steam engines	1800	Cornish roll crushers invented to grind ores in Cornwall, England
	1827	Portland cement produced
	1831	Bickford miners' safety fuse invented for blasting with black powder in Cornwall
	1838	Singer steam churn drill
	1839	Iron roller mills for grinding grains
	1840	Patent for hammer mill
	1847	Nitroglycerin introduced as first high-energy explosive
	1848	William Easby's high-pressure, double-roll compactor
	1849	Mechanical percussion drill
	1850	Fine sieves and aspirators for flour sifting and purification
	1858	Eli Blake jaw crusher
	1860	Rock drill powered by compressed air
	1867	Dynamite made from nitroglycerin
	1867	Fulminate of mercury blasting cap invented for nitroglycerin
	1867	Peter von Rittinger's energy–size reduction theory (first theory)
	1870	Rotary kiln to produce cement clinker
	1870	Alsing batch ball mill for grinding flint for potteries
	1871	Tripod-mounted rock drills
	1876	Gebruder Sachsenberg's screen-discharge ball mill
	1880	Corrugated iron rolls used for grain milling, leading to gradual reduction system
	1881	Philetus Gates' gyratory crusher
	1885	Askham dry air separator
	1885	Friedrich Kick's energy–size reduction theory (second theory)
	1888	Vertical roller mill (F.E. Huntington's mill)
	1889	Cyanide process for gold extraction
	1891	First patent on hydrocyclones
	1895	Electric delay blasting caps invented by H. Julius Smith
	1895	Thomas Edison's giant rolls
	1895	Water-flushed hammer drills
	1898	Ball mills used to grind gold ores in South Africa, Australia
	1898	Medium-speed pin mill
		Electricity
Electrical motors	1900	Vertical ring-roller and ring-ball mills for coal and soft ores
	1904	J.V.N. Dorr's mechanical hydraulic classifier
	1905	Flotation process for sulfide ores
	1906	Harlowe Hardinge's conical mill
	1907	Pebble milling used on the Witwatersrand (the Rand), South Africa
	1909	Cryogenic breakage

TABLE 1.6 Full chronology of size-reduction technology (continued)

Source of Power	Time Frame*	Machines and Processes
	1910	Rod mills
	1910	Raymond ring-roller mill (fixed ring)
	1920	Multiple-compartment tube mills
	1928	Loesche GmbH's roller mill (rotating bowl)
	1928	Vertical stirred ball mill
	1930	Will Symons' cone crusher
	1930	Vibrating ball mill
	1932	Alvah Hadsel's wet primary autogenous mills
	1933	Rosin-Rammler equation to define size distributions
	1935	Tromp size separation curves
	1941	Hardinge's "electric ear" for feed control
	1942	Dutch State Mines' cyclone for coal circuits
	1944	Humphrey Investment Company's cyclone with vortex finder
	1948	Hydrocyclones for ore circuits
	1950	Tungsten carbide drill bits
	1950	Jet mill
	1951	Fred Bond's energy–size reduction theory (Third Theory)
	1952	Delay detonators
	1955	Kubota Company's tower ball mill
	1956	Kelly Krebs' hydrocyclones
	1956	Broadbent–Callcott breakage-selection-classification matrix model for grinding mills
	1957	Ammonium nitrate slurry explosives
	1958	Jack Williamson's load-power-feed rate controller for pebble mills
	1960	Second-generation air separators with external cyclones
	1960	Down-hole hammer drills
	1963	Vertical high-speed "John" peg mill 205
	1970	Vertical-shaft impact crushers
	1975	Centrifugal mills
	1975	Computer-control techniques for grinding circuits
	1976	Nonelectric delay caps
	1977	High-pressure grinding rolls
	1980	Third-generation high-efficiency air separators
	1980	Computer-simulation techniques for mineral grinding circuits
	1993	High-speed horizontal disk mill
	1995	Very large roller and tumbling mills for minerals (~1,000 tph)

* Most dates are indicative.

THE PREVALENCE OF SIZE REDUCTION

Size reduction is not only the most ancient technology, it is also the most widespread. It is used in every country and in every industry that involves solid particles. Here are some of the more important applications of the technologies we profile in this book:

- Breaking wheat grains to separate endosperm from bran and grinding the endosperm to produce flour
- Grinding many foodstuffs, including spices, confectionery, and animal foods
- Breaking rock masses into boulders and pebbles
- Crushing and grinding ores to produce copper, iron, gold, and other metals

- Pulverizing coal to combust in power stations
- Grinding limestone and clinker to make cement
- Grinding industrial minerals to use as pigments and to manufacture fertilizers, glass, chemicals, pharmaceuticals, and many other products

In the next chapter, we discuss the contributions of science and scientists to the field of size reduction. Chapters 3 through 10 give details of the histories of the individual processes, and Chapter 11 takes a look at the future of size reduction.

CHAPTER 2

The Science and the Scientists

THE ROLE OF SCIENCE

Scientists and inventors have long worked to develop and improve size-reduction processes and machines to solve the engineering challenges associated with grinding. Size reduction—used in every mineral-processing operation—has been a continuing field for scientific inquiry and this has contributed to more efficient size-reduction processes in a number of areas:

- The design of better machines
- Modeling and simulation techniques to optimize circuits
- Control techniques to ensure that products meet tight specifications
- Laboratory techniques to measure material and process characteristics

Although little is known about the people who invented manual- and water-powered querns, black powder, or the stamp mill, they must have been excellent scientists, because their inventions solved the problems of their times. By 1700, the value of science to technological innovations and improvements was starting to be recognized. At that time, apprenticeships (or their equivalents) were the main form of training. Although excellent for teaching current technological practices, these types of arrangements were less successful at encouraging the search for new ideas. And, in the mineral industry, new ideas were urgently required because the growing demand for metals and ores was leading to

- Ore body depletion and loss (at the time, mining was done by following the veins that were exposed as mining proceeded, not by exploring ahead of mining using drill holes); deepening mines; and declines in metal grades
- The need to concentrate minerals with fine grain sizes
- Many accidents and deaths resulting from rock falls and poor access to deep mines

To alleviate a shortage of technically skilled workers who could better control and improve operations, and to protect the large revenues that governments received from mining operations, mining academies for advanced training began to be established. Universities had been offering advanced studies in philosophical and theological areas for hundreds of years, but engineering had been neglected. The rising interest in mining academies was a good indication that the link between prosperity and technical skill was being recognized.

Mining academies were established in Germany at Freiberg in 1766, Schemnitz in 1770, and Berlin in 1770; in Russia at St. Petersburg in 1773; and in Mexico City in 1791. These academies laid the foundations for the scientific advances in mineral engineering that occurred in the late 19th and early 20th centuries. The curriculum in these early academies mainly dealt with mineralogy, milling, and smelting.

The scientific culture moved easily into industry, although technology transfer must have seemed slow at times. The wry comment made by a distinguished smelting metallurgist who trained at Freiberg is pertinent: "[Up to 1860] the percentage recovery in the metal was practically altogether neglected, and smelters worked and were happy in a sort of metallurgical dreamland, trustingly hoping that whatever was right" (Sticht 1905).

By 1900, finer grinding was in great demand, particularly for the portland cement, cyanide, and flotation processes. The importance of science in improving processes was becoming very clear. Because breaking more particles or making a finer product requires more energy, the starting point for scientific improvements in grinding was to understand how the product size changed when more energy was used. A key question emerged: What is the size of mill required to grind ore to a required product size if the feed rate and grindability are known? Devising a good test for grindability has always been only part of the answer; the other part relates to accurately predicting the power required and then choosing the size of the mill that will absorb this power.

Peter von Rittinger and Friedrich Kick published two conflicting theories on this area of inquiry—the energy–size reduction relationship—during the last half of the 19th century, and these generated much debate and research. Missing from the original work and the later controversy was an accurate method of measuring the size and surface area of the dust particles that were inevitably generated and that formed most of the surface area. The controversy could not be resolved until new measurement techniques became available, and by the time these were developed, the question was of academic interest only because many other factors had come into play by then. But the controversy was useful, because it encouraged engineers to see the necessity of measuring process variables and to think about processes in quantitative terms. Research on the energy–size reduction link, which continued without pause, led to a third theory by Fred Bond in the mid-20th century. Bond's theory became widely used for plant design.

By 1910, tumbling mills were coming into their own. The development of tumbling mills is discussed in Chapter 7, but we should mention a vital link here: The demand for these mills coincided with the rise of electrical power.

During the 20th century, electricity improved the technology of size reduction from fledgling machines and low-capacity processes to huge machines and immense-capacity processes, changing the mining industry dramatically by 1920. By 1990, the personal computer had become the easy-to-use means to make the necessary mathematical calculations to simulate the performance of size-reduction machinery and circuits, and modeling and simulation was becoming an essential tool for design and operation. By 2000, engineers were building motors so powerful that they no longer limited machine size, and scientists were developing blasting procedures using huge drills and new explosives that could shatter any quantity of rock in a single blast. The focus had changed from searching for power to ensuring that power was used as efficiently as possible.

Science was a crucial factor in all these impressive developments. Of course, in an inherently human feature of progress, not all changes proceeded smoothly. It was often difficult to create enthusiasm for change. In 1890, the great copper mining region in the United States known as the Lake Superior District was typical:

> *The real reason for want of progress in the concentration of ores in Lake Superior is that everyone is doing the same thing and no one is willing to take the first step to advance. Some of the mills are experimenting in a vague sort of way but all of the small mines expect to reap the benefit without cost of the experiments which the large mines ought to make and the large ones expect the largest ones to take the first step and find the paying improvement. (Benedict 1955)*

This attitude changed in time. By 1915, for example, the potential value of ball mills was recognized at the Calumet and Hecla mine in Michigan's Upper Peninsula when, even though the concept of ball millings was still unproven, 64 mills were installed in a new tailings retreatment plant. The plant operated successfully and produced large volumes of copper concentrate for years.

In the rest of this chapter, we introduce some of the scientists whose names have gone down in size-reduction history. Many others made significant contributions to the field, but we mention only a few here to give a flavor of how science and technology advanced in tandem to improve size-reduction processes and machines.

THE SCIENTISTS AND THEIR CONTRIBUTIONS

Peter Ritter von Rittinger

Peter Ritter von Rittinger was an outstanding mineral scientist whose equipment inventions and mathematical descriptions of unit operations established mineral processing as a modern technology (Steiner 2000; Vozar and Kunnert 1972). He was the first to provide a sound mathematical basis for the physics of wet classification and gravity separation processes, and the first to discuss the quantitative relationship between the energy consumed and the size reduction achieved in crushing and grinding.

Von Rittinger was born in 1811 in Neutitschein/Mahren, Austria, to parents who died when he was very young. His education was financed by scholarships. After studying philosophy and jurisprudence at the University of Olmutz, Austria, he went to the Austrian Mining Academy in Schemnitz, where he graduated with honors as a mining engineer in 1839. He then worked in various capacities with state-owned mining enterprises. He was assigned to supervise and modernize mineral processing plants, which in those days were mainly based on grinding and gravity separation circuits. Separation processes of the time were batch rather than continuous, and von Rittinger saw the need for change. His first two inventions were the Spitzkasten, which enabled classification of fine particles ahead of gravity concentration, and the continuous shaking table, which was the basis of the later and more successful Wilfley table.

In those days mining engineering included mining and mineral processing, and von Rittinger moved around the industry, taking on a variety of tasks in both areas. Selected highlights of his career included serving as a coal exploration officer at Brandeisel, where he was responsible for the rapid sinking of two shafts; managing mining at Joachimsthal, where he designed the deepest shaft of the time; designing ventilators for mines, a concept that was used for 60 years; and inventing the thermocompression process for the evaporation of water from salt brines.

Von Rittinger earned a worldwide reputation as an outstanding scientist when he published a comprehensive textbook on mineral processing (*Lehrbuch der Aufbereitungskunde*; von Rittinger 1867). The book described equipment and processes, and was accompanied by two heavy volumes of beautiful lithographic drawings of all the technical equipment that operated in mineral processing plants at the time. But his book went much further than describing machines and processes. He recognized the importance of quantitatively explaining the principles involved in the processes, using equations to do so. Between his inventions and his outstanding textbook, it is not surprising that von Rittinger is regarded as the man who laid the foundation for today's mineral processing technology. His achievements were of such stature that he was awarded civic as well as technical honors, including being ennobled by the state.

In discussing size reduction in his textbook, von Rittinger commented that "The throughput (of wet operated stamp mills) is proportional to the 0.4 power of the linear dimension of the openings of the discharge screen." This type of empirical relationship, which was unusual at the time, was similar to the later Bond equation. To clarify, he used an analogy of making estimates of the new surface produced by repeated breakage of particles into smaller particles. Although his original equation was correct, the analogy suffered the fate of many simplifications when other authors took it to be an actual hypothesis (Steiner 2002). Eventually the energy–size reduction relationship attributed to von Rittinger came to be stated in this way: "The work done in crushing is proportional to the area of new surface produced" (i.e., proportional to the reduction in linear dimension; Richards and Locke 1940). The hypothesis attributed to von Rittinger is defined by the equation $E = k[1/x_2 - 1/x_1]$, where E is the energy, k is a constant, and x_1 and x_2 are the feed and product sizes. These symbols have the same meaning for the Kick and Bond equations.

Friedrich Kick

Friedrich Kick was born in Vienna in 1840. After graduating from the Vienna Polytechnikum in 1862, he became professor of mechanical technology at the Technical University in Prague, Bohemia, where he worked from 1865 to 1892, including three terms as rector. From 1892 to 1910, he was a professor at the Technical University in Vienna (serving as rector there in 1895). Kick was politically active and served in the regional parliament of lower Austria (Steiner 2002). He became well-known for his experiments on material properties and his development of test equipment to measure these properties. His publications, especially the 1885 book, *Law of Proportional Resistances*, added to his reputation.

Kick's experiments that led to his hypothesis on the link between energy and size reduction started with studies of the deformation of soft bodies under load. He concluded that the amount of deformation is proportional to the energy applied and that "The pressure required for similar deformation of similar shaped bodies of similar material is proportional to the cross section area of those bodies" (Kick 1883).

This work was extended to the breakage of stone, iron and glass spheres, and cylinders, and then to rock breakage by blasting. The result of these studies was the energy–size reduction hypothesis attributed to Kick: "For any unit weight of ore particles the energy required to produce any given reduction ratio in the volumes is constant no matter what may be the original size of the particles" (Richards and Locke 1940). Kick's hypothesis is defined by the equation $E = k[x_1/x_2]$.

By 1915, a scientific culture was well established in the mineral industry, and technical publications of the day indicated that engineers had developed skills in measuring and improving the performance of grinding circuits (Del Mar 1917). The scene was set for a better understanding of the energy–size reduction relationship.

Fred Bond

Fred Bond was the engineer who did much to define the relationship between ore hardness, tonnage processed, size reduction achieved, and power required. Born in 1899 and raised in a rural community near Golden, Colorado, he attended and graduated with honors from the Colorado School of Mines (CSM) with a bachelor's degree in metallurgy. He went to work as a metallurgist for New York & Honduras Rosario Mining Company at its mines in Honduras, where he learned about the pebble milling studies done there in the early part of the 20th century. He later returned to CSM for his master's degree in metallurgy, then went to work at the U.S. Mint in Denver where he assayed the gold shipments. In 1928, he became a draftsman for the Tennessee Copper Company in Copperhill, Tennessee, where he worked with Mill Superintendent Jack Myers. Myers was a student of grinding and worked closely with Harlowe Hardinge, who was conducting research on grinding in the mills he had sold to Tennessee Copper.

A casualty of the economic downturn during the Great Depression, Bond was released by Tennessee Copper in 1930. Walter Maxson, who had been one of Bond's professors at CSM, was managing the Mining Department at Allis-Chalmers, which sold

THE RITTINGER–KICK CONTROVERSY

The Rittinger–Kick controversy was prominent in the technical press from about 1900 to 1940. Much scientific thought was expended and many laboratory tests were made, but there were no definite conclusions. Some of the data indicated that Kick agreed better with crushing and Rittinger with grinding. . . .After about 1940 the controversy abated as being nonproductive. (SME–AIME 1985)

In a simple way the difference between the theories can be shown by considering the breakage of a cube of rock with 1-m sides. Assume that the original cube is broken into cubes with sides of 0.5 m, then 0.25 m, 0.125 m, and so on. After each breakage the volume of the cubes is one-eighth of the previous volume (i.e., the ratio of the volumes is constant), but the surface area doubles, from 6 to 12 to 24 units, and so on. Rittinger's theory posited that the energy increased as the particle size decreased and surface area increased, but Kick's theory postulated that the energy required for each step was constant. Experimental studies tended to support Rittinger's theory, although Kick's theory seemed to apply to coarser particles.

gold ore–processing systems at the time. Maxson hired Bond to design and build a grinding and metallurgical testing laboratory. In the 1930s, Bond traveled to Peru and Bolivia to start up gold ore–processing plants, and during World War II he was involved in the start-up of a radium ore mine at Great Slave Lake in northern Canada. All the while he continued his studies on grinding, and eventually he developed the rod and ball mill grindability tests that now bear his name. The method Bond developed for selecting ball mills (see sidebar) is still routinely used by mineral process design engineers.

Bond and Grindability Throughout the 20th century, grinding mills were selected based on the grindability of the material and the prediction of the power required per ton to grind the material to a known product size. Before Bond's work, this estimate was based on experience and judgment. Metallurgists and process engineers who worked with mill manufacturers traveled extensively and gained wide exposure to plant data. Because it was generally accepted that they had a better knowledge of the relationship between grindability data and actual mill performance than the staff of mineral processing companies, whose experience was necessarily restricted to a few operations, engineers from mineral processing companies consulted with representatives of grinding mill manufacturers when mills and circuits were being selected for new mines and plant expansions.

To compete for the grinding mill business, manufacturers needed to increase their expertise in grinding technology, so they developed their own grindability tests and obtained operating data and corresponding ore samples from processing companies. From these, they could derive the relationship between grindability and mill performance. The early grindability tests were batch tests carried out in small-diameter ball or pebble mills, often called jar mills. The results were given in terms of revolutions of the test mill or time required to make the desired particle-size distribution and net weight of a specified mesh size produced per minute or per set number of revolutions in the test mill.

Each business that developed grindability tests also developed its own testing procedures, which were kept confidential. Both wet and dry test procedures were developed, but dry grinding was found to give more reliable and reproducible data. During the 1930s, instruments and techniques were developed to measure grindabilities, although the data from the tests were generally not used to directly select grinding energy. Instead, engineers searched for another, usually similar material that had the same grindability as the test material and for which plant production data were available; they then used these data to predict the energy required to grind the test material. The accuracy of the energy prediction depended on the knowledge and experience of the engineer making the selection.

Bond and the Work Index Equation The combination of physical data, particularly plant operating data recorded in the engineers' "black books," and the knowledge of how to use those data were important. The "old timers" carefully guarded their black books and shared as little as possible with associates as a protection against newcomers eager to take their jobs. Often they died or retired without disclosing how to use the black books, which rendered the books of little use. Bond's research started a new era in grinding circuit design that did not involve looking for comparable operating data, and the need for the black books disappeared. The first half of the 20th century can be called the pre-Bond era, and the last half can be called the Bond era. The Bond grindability test procedure was completed in 1937.

During the late 1930s and 1940s, Bond devoted as much time as possible to understanding the energy–size reduction relationship. He began to develop the concept that the energy needed for grinding was the total energy needed to make the grinding mill product minus the energy needed to make the feed. Few companies kept accurate power data for their mills at the time, and the available data were in terms of amps, volts, and nameplate information on the rating of the motor. There was no power factor information to calculate motor power draw, so each mill manufacturer worked out a method to determine the mill power for designing mill drives.

Bond's work in this area led to the concept of the "work index," which is the kilowatt-hour per short ton required to reduce material from theoretically infinite feed size to 80% passing 100 μm. In establishing the work index equation, Bond assumed that the efficiency of classifiers was consistent, and he did not consider this to be a factor in determining the energy needed for grinding. In 1952, the Third Theory of Comminution, which contained the work index equation $E = [1/\sqrt{x_2} - 1/\sqrt{x_1}]$, was published (Bond 1952). This equation related power required per ton to feed and product size and grindability. Bond's method for determining grinding energy has become the standard. It is now used universally to quantitatively define the resistance of solid particles to breakage and is employed for grinding circuit design over a wide range of operating conditions. Although correction factors have been introduced, the procedure for design has remained unchanged. The work index equation is unlikely to be supplanted, although simulation techniques will augment it.

In the 1950s, demand was increasing for larger mills from both the mining and cement industries. Bond used the data he had collected to derive new power draw equations for rod and ball mills, and he was permitted to publish these in *Crushing and Grinding Calculations* (Bond 1961). There were small differences between manufacturers in how they rated the power draw for rod and ball mills, and this led to the rating of rod and ball mills by the power they were designed to draw.

Bond's "computer" was a log-log slide rule that was 500 mm long and had slightly larger numbers than were standard on slide rules at that time. The accuracy of the slide rule was a limiting factor in the accuracy of calculations, and Bond's larger slide rule improved the accuracy. He concluded that, if he plotted the natural size distributions of many materials on log-log plots, the sections of the graphs between the 80% and 20% passing sizes were essentially straight lines. Beyond these limits the graphs were irregular. So he selected the 80% passing size in microns as the most accurate and useful value to describe a size distribution. Sieves were used to give perfect classification in the closed-circuit grindability tests.

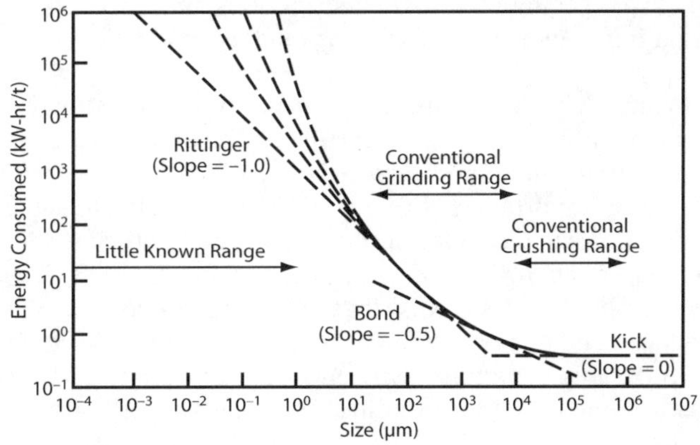

FIGURE 2.1 Relationship between energy and particle size in breakage (Hukki 1961)

Finland's R.T. Hukki reviewed data from many industrial operations and concluded that the three energy–size reduction theories referred to different regions on the curve relating energy used to particle size produced, as shown in Figure 2.1 (Hukki 1961).

This graph makes it clear why Bond's work was so successful and why it will continue to be used to design circuits that grind to about 200 μm. His work covered a product range of 25,000 to 20 μm (25 mm to 0.02 mm), and most applications will continue to fall in this range. But extending this work to products less than 5 μm is incomplete.

After Bond retired in 1964, engineers at Allis-Chalmers continued to develop the Bond equation and publish their results. The book *Design and Installation of Comminution Circuits,* published by the Society of Mining Engineers in 1982, was dedicated to Bond, who died in 1976.

Paul Rosin and Erich Rammler

In the days before digital computers, sizing distributions were often simplified to a single value, typically the size in microns through which 50% or 80% of the material passed. This was good enough for Bond's work on circuit design but not good enough for understanding breakage mechanisms or identifying how problems in breakage occurred. Complete size distributions were needed, and it was Paul Rosin and Erich Rammler in Germany who showed that sieve analyses of pulverized coal could be fitted by a simple negative exponential equation.

> *During World War I the combustion of hard coal dust had been developed in the USA and put into large scale operation. After the war the industrial countries of Europe followed suit. In Germany the mining of brown coal and hard coal had been greatly stepped up. Efforts were therefore made to introduce the combustion of brown coal dust. In this connection Dr-Ing Paul Rosin, Head of the Department of Heat Economy of the Staatlichen Blaufarbenwerke in Freiberg and unsalaried lecturer at the Bergakademie Freiberg, set himself the task of investigating the special problems of the combustion of brown coal dust, in particular for metallurgical furnaces. (Schubert and Wachtler 1987)*

Rosin started his research work on grinding and combustion processes in 1923, at the same time that Curt von Grueber was installing Maxecon mills in the Moabit power station in Berlin. Rosin set up several mills and air classifiers in a pilot plant in a smelter near Freiberg in an effort to understand how fine particles were generated during size reduction and to develop equations of size distributions. In March 1925, after graduating from the Freiberg Mining Academy with an outstanding academic record, Erich Rammler joined Rosin.

In 1925, construction began on the first brown coal power station in Germany, which was near Leipzig and Rosin's experimental plant. Inevitable start-up problems were encountered because unwanted fines were present when the power station came on-line, which emphasized the importance of fine particle studies. Rosin set up a program in which feed and product size distributions were measured and changes in the operating variables were related to changes in the size distributions. In 1933, Rosin and Rammler published the equation for the size distribution of broken coal: $x = 100.\exp(-b.x^n)$, where x is the oversize and b, n are variables that describe particle characteristics (Rosin and Rammler 1933). This equation has been found to be applicable to many broken products and has become a tool for grinding studies, even being used for assessing the efficiency of dentures.

Rosin and Rammler extended their brown coal research to drying and firing and became consultants in brown coal processing. In 1937, Rosin emigrated to England, and, in 1945, Rammler became director of a research program on briquetting and degasifying brown coal at the Bergakademie Freiberg, where he remained for many years.

Rosin-Rammler and John Bennett

Although John Bennett had a flair for mathematics at school, he did not take the scholarship he won to Oxford University. Instead, when World War I started, he joined the Royal Engineers at the age of 17. After the war he lived in Turkey and Greece for 15 years, where he worked as a diplomat and as an engineer on brown coal mining. Upon his return to England, he joined the British Coal Utilisation Research Association (BCURA), where he was appointed as its first director. BCURA was established in 1938, taking over current research on coal (McCaffrey 2003).

At the time, households or small industries that required graded or semigraded coal formed the main market for coal. Coal was broken from the seams by pick and shovel or blasting and prepared for market by screening into fractions. Fine coal burned poorly and this was a particular problem with the friable British coals. Hypotheses about maximizing the salable products were primitive at best. Even though the Rosin-Rammler equation was the first real clue about coal breakage, the exponential in the equation was difficult to handle, so for a time it went nowhere.

In 1936, Bennett, in collaboration with R.L. Brown (a mathematician from Oxford University), examined size analyses of run-of-mine coals from many British mines. These analyses were well fitted by a revised Rosin-Rammler equation: $x = 100.\exp(-(x/a)^n)$, where a is the size at which (100/e) = 36.8% of particles are retained (Brown 1941). This equation, which became known as the Rosin-Rammler-Bennett (R-R-B) equation, opened the door to relating changes in parameters in an equation to changes in coal types and breakage conditions.

In 1956, Tom Callcott and Simon Broadbent took this idea further when they proposed a model in which a breakage function based on the R-R-B equation defined the coal type and the probabilities of breakage and discharge defined the breakage conditions. To simplify calculations, they wrote the model in matrix form, and it gave insight into how particles were broken into size fractions (Broadbent and Callcott 1956). This model proved to be suitable as a basic engineering model of size-reduction machines, and from it has come some of the comprehensive models of machines that are now so important in optimizing size-reduction processes.

Similar models were developed in the 1960s in other laboratories. The earlier models are discussed in *Crushing and Grinding: A Bibliography* (Bickle 1958).

Oliver Evans

Although the scientific techniques of observation, deduction, and experiment to improve size-reduction processes were in widespread use, there had been few if any case studies until the work of Oliver Evans. Born in Delaware in 1755, Evans became one of America's great pioneering inventors. Delaware was a good place for him to spend his formative years because of its proximity to Pennsylvania and Virginia, where there was much intellectual and industrial activity associated with the growing population. His first invention in 1777 was a machine to produce teeth for wool-combing cards and came about because he disliked making them by hand (Lienhard 1988–1997). In 1781, he joined two of his brothers to buy a flour mill and soon discovered its deficiencies. Flour milling at the time involved taking batches of cereal grains through a series of processes and holding some of the products in piles for up to several hours before further processing. It was slow, cumbersome, labor-intensive, and rather unhygienic because stockpiles of partially processed flour attracted insects.

In 1782, Evans devised a system to change flour milling to a fast automatic process in which the stockpiles were eliminated. The flour mill he built on Red Clay Creek near Newport, Delaware, used a system of shafts, gears, and belts to drive the machinery in the plant from a water wheel, along with elevators, conveyors, and gravity pipes to move the partially processed flour from one machine to the next. Hoppers were used during sifting and for drying the grain. This system was very successful, and the book in which he described it passed through 14 editions (Evans 1785, 1805).

Yet flour milling does not seem to have been his main interest. Starting in 1784, he patented steam engines—notably a high-pressure engine—which were used in locomotives and factories. These engines made enormous contributions to America's rapid industrial growth in the early 19th century. His book on steam engines, which contained a "description of a steam engine on new principles, rendering it much more powerful, more simple, less expensive, and requiring much less fuel than an engine on the old construction," was a fine scientific treatise (Evans 1805).

John Van Nostrand Dorr

Born in New Jersey in 1872, John Van Nostrand Dorr started work as a laboratory assistant in Thomas Edison's laboratory in Orange, New Jersey, at the age of 16. At the time Edison had become heavily involved in mining and processing low-grade magnetite ores. Thinking that the iron ore deposits in the eastern United States were being exhausted, he saw an opportunity to enter the market for iron ore. The mining and concentration processes of the day were small in scale and expensive, and new high-capacity processes were essential to the economic success of the ventures. Edison established a laboratory to provide data and information about the processes he was planning.

Dorr worked in the laboratory for a little more than 2 years. During his last year there, he worked directly for Edison. He must have absorbed much of Edison's innovative approach to identifying and solving problems, because Edison's slogan of "try anything once" was evident in many of Dorr's achievements. After leaving the laboratory he studied chemistry at Rutgers University and worked at a gold mine in South Dakota one summer. The mine was not profitable because the ore consisted of a clay fraction with no gold and a sands fraction containing gold, and the clay interfered with the extraction of gold. Dorr learned this lesson in the limitations of mineral processing technology well, realizing that if the clay could be separated from the sands and only the sands processed, the operation would be profitable.

After graduation and a brief term as a research assistant at Brooklyn Polytechnic Institute, Dorr returned to South Dakota. At the time, cyanide leaching was being introduced to gold mines in the United States. In 1903, Dorr took part ownership of the

Lundberg, Dorr and Wilson mill that was built near Terry, South Dakota. In 1904, he tested the idea of a sand-slime split using a classifier with reciprocating rakes to replace the classifying cones. It operated by keeping the finer particles in suspension through the stirring induced by the rakes and overflowing them into a launder. The coarser particles fell to the base of the machine through the force of gravity and were continuously removed by being raked into another launder. The Dorr rake classifier gave a coarser and more controllable split than was previously available with cone classifiers. Used for the first time in 1909 in closed grinding circuits at the Real Del Monte mine in Pachuca, Mexico, the Dorr rake classifier was soon used in many circuits.

It is said that when the classifier was built and was first operating, Dorr worked or slept beside it for its first 80 hours of operation. His brother G.H. Dorr wrote that J.V.N. Dorr's

> *...ability to work indefinitely on 4 or 5 hours sleep and an instinct for starting a session for the concentrated consideration of some problem at some odd hour of the day or night is one of those maddening characteristics for which his associates were at times inclined to curse his early association with Edison. (Dorr 1941)*

Dorr also invented the continuous thickener in 1906, which was the first continuous sedimentation unit in which water could be removed from a dilute slurry and the solids concentrated into a dense slurry. The continuous thickener allowed water to be recycled from the mill tailings pond to points within the mill circuit, greatly reducing the volume of fresh water that had to be used in the mill.

Mineral companies quickly accepted both the rake classifier and the thickener. Wet closed-circuit grinding became part of mineral processing technology and the thickener became essential for recycling process water, which was often very scarce.

Eli Whitney Blake

Eli Whitney Blake was a member of a leading family of industrialists in New Haven, Connecticut, which had a tradition of invention and entrepreneurship. The family's main business was the manufacture of arms, but family members participated in civic affairs and assisted in planning the development and servicing of the city. Genetics may have been a factor in Blake's inventiveness–his mother was the sister of Eli Whitney, who invented the roller cotton gin.

In 1852, the town council decided to pave Whalley Street, a 2-mile-long thoroughfare in the central part of the city. Blake was one of the townsmen responsible for planning the paving project, which employed the MacAdam system–in use for road building since 1815. In this system, the surface of a road was formed by interlocking stones that had passed through 2-in. apertures and then sealed. The practice of the day was to break rocks by hand, which was slow and costly. Realizing that a machine could break quarried rock faster and cheaper than a man, Blake designed the first efficient jaw crusher (Dickinson 1945). This invention was remarkable; he used converging fixed and moving jaws and drove the moving jaw through a toggle joint that exerted immense pressure on the jaw when its members were moved into a straight line. The design, which was prepared before the first machine was built, has changed little in its essentials for 150 years. Jaw crushers built to Blake's design are still widely used today.

Hans Rumpf

Because the breakage environment is so different between laboratory and plant, size reduction is not a subject that readily lends itself to the classic technique of conducting basic research and then transferring the results to large-scale operations. But Professor

Dr.-Ing Hans Rumpf showed that the links between empirical and fundamental knowledge could be understood through research, and the work of his research group contributed a great deal to the more efficient use of energy in size reduction.

Rumpf was born in Langenschwalbach, Germany, in 1911 and graduated from the Technical University of Karlsruhe in 1939. From 1936 to 1942 he worked in the dye industry, then for Alpine AG in Augsburg, Germany, from 1942 to 1955, where he designed new machines to grind and classify fine particles. This background gave him the experience needed to direct a new fine-particle research institute at the University of Karlsruhe, where in 1956 he became a professor of mechanical process technology. From 1966 to 1968, he was rector of the university, and, from 1968 to 1971, he served as president of the West German Rector Conference. A measure of the success of the research institute is that, by 1970, it had a staff of 116, including 45 scientists, and graduated 160 students with diplomas and 30 with doctoral degrees. The institute also generated 225 publications.

Comminution science was a main area of research in the laboratory, and studies were carried out on

- Fracture physics
- Crushing under controlled conditions with special reference to single-particle crushing, measurement of energy needed for fracture, fragment size distribution, and new surface area produced
- Mathematical simulation of comminution processes
- Practical comminution

In 1976, Rumpf reviewed some of the laboratory's work in a lecture to the Fine Particle Society in England (Rumpf 1977). He commented in the lecture that

> *In comminution the particles will react individually in a greater measure than in any other process. The reason for this is that the defect and dislocation structure is decisive for the fracture formation. As a result the breaking behaviour will, for example, vary to a great extent with the kind of material and with particle size but can also vary greatly with the same material and particle size.... How much energy can, in fact, be saved can only be determined by systematic tests with single particles.*

Rumpf compared the specific energy consumed by materials crushed as single particles and broken in industrial comminution processes. His results are shown in Figure 2.2, which indicate that energy utilization could be improved by a factor of 10 in theory (although it would be less in practice). One of the outcomes of Rumpf's work was the development of high-pressure grinding rolls (HPGRs), which was one of the main successes in new grinding equipment during the 20th century.

Klaus Schonert

Dr.-Ing Klaus Schonert studied under Rumpf at the University of Karlsruhe, where he used single-particle breakage to study comminution, studied air separators to close dry-grinding circuits in cement plants, and did his early studies on HPGRs. In 1968, he became the head of the university's comminution group (Instituterfuer Aufberetung und Veredelung). Schonert's research at the University of Karlsruhe and later at the Technical University in Clausthal, Germany (Technische Universitat Clausthal), led to the patenting of the HPGR in 1982. He is best known for his work in developing the high-pressure double-roll crushing and grinding mill used for coarse grinding of cement clinker.

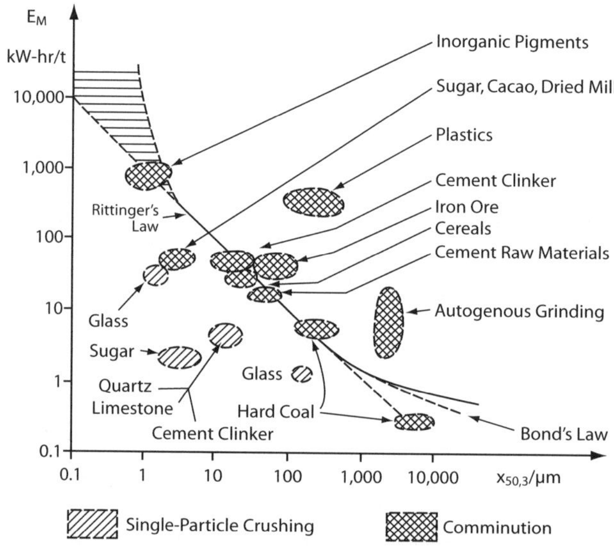

FIGURE 2.2 Specific energy consumption in comminution and single-particle crushing (Rumpf 1977; reprinted by permission from Elsevier Power Technology)

Immanuel Nobel

Immanuel Nobel, Alfred's father, was a civil engineer and inventor in Stockholm, Sweden. Immanuel's interest in discovering a powerful explosive suitable for breaking rocks started with the excavations involved in his construction work and was fueled by plans to cut a canal at Suez, Egypt.

> *It was incredible how little progress had been made in the ways of canal and road building since the time of the Roman Emperors. ...Mining especially needed means by which to cut faster into the subsoil. A powerful explosive that was capable of blowing up rocks at vital spots could revive the only industry worthy of the name in Sweden–the mining of iron ore. (Halasz 1959)*

In 1837, Immanuel Nobel moved to St. Petersburg, where he convinced the Tsar of Russia that naval mines could protect the port from foreign warships. The resulting contract enabled him to educate his sons well and to finance Alfred's 2-year visit to Sweden, Germany, France, and the United States in 1849–1850. In France, Alfred met Ascanio Sobrero, who had invented nitroglycerine 3 years earlier. This meeting had great consequences.

Although nitroglycerine is a very powerful explosive, its sensitivity to heat and pressure made it too dangerous to use. In St. Petersburg and then in Stockholm (to which they returned in 1863) Alfred and Immanuel systematically investigated the possibilities for making nitroglycerine safer without losing its explosive power. Their experiments involved detonating mixtures of nitroglycerine and black powder under various conditions. They concluded that the full effect of nitroglycerine as an explosive could only be obtained by detonating it with a compound, such as mercury fulminate, exploded in close proximity (Drinker 1888), and the igniter cap was born. Although the market for nitroglycerine explosives was brisk, premature explosions still occurred.

In 1867, Alfred observed that the mineral kieselguhr, which was packed around containers of liquid nitroglycerine, absorbed the explosive when it leaked. He made a mixture of 75 parts of nitroglycerine and 25 parts of kieselguhr that was solid, could be formed into cartridges for putting into drill holes, was less sensitive than nitroglycerine, and had strong explosive properties (McAdam and Westwater 1958). He called the product dynamite and demonstrated that it could be handled safely and detonated when required. With dynamite, Nobel took blasting to another plateau of safety, efficiency, and cost, and this explosive dominated the market for 100 years. With this discovery, "The blasting of roads and mines now entered a new era, as did the military art" (Halasz 1959).

Later Nobel combined guncotton (made in 1846 by Christian Schoenbein at the University of Basle, Switzerland, by reacting nitric and sulfuric acids with cotton) with nitroglycerine to make a very powerful explosive that he called blasting gelatin. He continued experimenting and improving his explosives while building a network of manufacturing companies and marketing offices in many countries. By applying science to the formulation of explosives, he made products that changed the world. The Nobel Foundation's Web site sums up his contributions well: "Alfred Nobel's greatness lay in his ability to combine the penetrating mind of a scientist and inventor with the forward-looking dynamism of the industrialist" (Nobel Foundation 2004).

Melvin Cook

The advantage of dynamite was the high energy it delivered into the rocks to break them; its disadvantage was that it was dangerous to use. Melvin Cook, one of the great mining scientists and entrepreneurs of the 20th century, invented a range of safer, easier to handle, and easier to transport explosives based on ammonium nitrate and fuel oil. These explosives, which were less likely to explode spontaneously and delivered much more energy, soon displaced dynamite as the dominant explosive.

After graduating from Yale University in 1937 with a doctorate in physical chemistry, Cook joined DuPont and worked on commercial explosives until the United States entered World War II, when his research was directed toward military explosives. After the war he was appointed professor of metallurgy at the University of Utah, and for a time his research centered on flotation theory. In 1952, the U.S. Army set up an explosives research group at the university, naming Cook as its director. His subsequent research focused on the deflagrating explosive ammonium nitrate, which had been used for blasting rocks for several years.

Because of ammonium nitrate's solubility in water, it had to be packed in a can for use, but Cook took an unusual approach that turned its solubility into an asset rather than a liability. He developed water-based explosives with a very high ammonium nitrate content. The ammonium nitrate–aluminum powder–water explosive he used at the Iron Ore Company of Canada in 1956 proved to be safe and efficient, starting a change in explosives technology that led to mixing explosives on site and pumping them into holes of any inclination.

Like Nobel, Cook took a leading role in transferring this new technology to industry by establishing consulting and manufacturing companies. Throughout his distinguished career, he retained his strong scientific interests, publishing six books on explosives and more than 200 papers in leading journals. Khodorovskiy (2000) reviewed Cook's life and achievements.

CONCLUSION

Again we emphasize that the people mentioned in this chapter are only a few of the many scientists who have applied scientific principles to industrial problems. Studies on size-reduction processes and machinery have been done in many different areas of the world by scientists representing many different entities, beginning with users of the technology and expanding to academia; to private, governmental, and commercial laboratories; and to machinery manufacturers. Without exception, the outstanding technical skills of these scientists and the dedication they applied to their tasks vastly improved the practice of humankind's oldest technology and set the stage for countless improvements.

CHAPTER 3

Hand Stones

His berries, nuts, acorns and corn were pounded into meal; and his invention of the first pounding appliance constituted, without doubt, one of the earliest attempts to devise an instrument useful in the peaceful arts. (Bennett and Elton 1898–1904)

In this chapter we discuss how stones driven by muscle power—which were first used for grains and later for ores, pigments, and drugs—advanced from the simple mortar and pestle (which produced coarse grains) to the saddlestone (used for much finer grinding) to the complex rotary quern (the first complete "machine").

MUSCLE POWER FOR BREAD

Since the beginning of humankind, the staple in our diet has been bread made with flour, which is produced by reducing the size of grains. Wheat grains consist of endosperm (~82%), germ (~2.5%), and bran (~15.5%). Flour production involves grinding grains to separate endosperm from bran, removing as much bran as possible, then grinding the endosperm to flour.

Originally grains were softened by soaking in water, and the entire softened grain was baked into bread. Legend has it that soaking led to brewing:

The Egyptians believed that one day Osiris, god of agriculture, made a decoction of barley that had germinated with the sacred waters of the Nile and then, distracted by other urgent business, left it out in the sun and forgot it. When he came back the mixture had fermented. He drank it and thought it so good that he let mankind profit by it. (Toussaint-Samat 1994)

Accidental fermentation may well have been the process preferred by workers during the hot summers, but it did not produce enough soft grain to eat, so the grains were ground with hand stones. Shear and compression forces break the husks away from the endosperm in cereal grains when particles are passed between a rotating surface and a fixed surface. For thousands of years, these forces were harnessed to make flour in family dwellings by grinding grain in small hand-operated mortars and pestles and with saddlestones.

In ancient Egypt the process was to crush the grain lightly in a mortar, separate the liberated bran by winnowing and coarse sieving, regrind the endosperm with a hand stone, and remove more bran by fine sieving with sieves made from hair or fine vines. The same process was also used in the Greek and Roman empires. Simple hand stones are shown in Figure 3.1.

Visitors to the underground cities in Cappadocia, Turkey, can see an example of a 2,000-year-old hand stone—pictured in Figure 3.2—used for grinding spice and foodstuffs. The underground cities were built during the course of hundreds of years by cutting rooms into the soft volcanic tuff. They consisted of tunnels, rooms, and ventilation

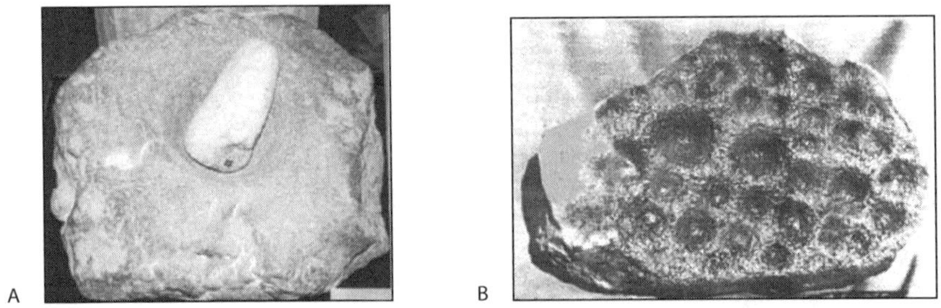

FIGURE 3.1 Hand stones used for grinding ores: (a) a conventional mortar and pestle (b) a multiholed mortar, an andesite rock; its weight is 25.6 kg, and its top area is 1,078 cm^2. In the top side, there are 33 holes, 42–77 mm in diameter and 8–20 mm deep, and in the bottom side, there are 9 holes, 56–90 mm in diameter and 9–28 mm deep (reprinted by permission from Maden Tetkik ve Arama Enstitusu Museum, Ankara, Turkey).

FIGURE 3.2 (a) A 12-ton grinding stone used for grinding spices and other food products in the underground city at Derinkuyu in Cappadocia, Turkey, more than 2,000 years ago. (b) A sketch of a section of the city. The communal grinding stone was on an upper level (adapted from Demir 1998).

FIGURE 3.3 The hopper-rubber, a grinding stone with converging sides and a slit in the base to control the product size. Materials were pushed through the slit by a stone.

FIGURE 3.4 A painting on a tomb in Egypt from about 1400 BC. After oxen break the husks from the kernels, the husks are rejected by winnowing (Casson 1966).

shafts, which were extended downward through many levels. During times of danger, theses underground cities served as residential quarters for thousands of people. The hand-grinding stone shown in Figure 3.2a was made from ignimbrite, a hard volcanic tuff, and is 2.5 m in diameter, 1 m thick, and 12 tons in weight. Its size made transport through the curved, narrow, sloping, dark passages of the underground cities extremely difficult.

Another type of hand stone used for breakage was the "hopper-rubber," a rectangular stone with a V cut into it. The V was wide at the top and converged to a narrow slit at the bottom. Nuts and grains were pounded and rubbed until the fragments were small enough to pass through the slit.

The photograph in Figure 3.3 shows one of many stones that were part of the cargo on a boat that sank in the eastern Mediterranean Sea about 300 BC. (The stones were recovered and are on display in a museum in Gyrenia in northern Cyprus.) There was a considerable trade in these stones, which were large, heavy, and resistant to abrasion. Because they were expensive, they were used with care so that they would last for many years. Amphoras, which were storage vessels used to transport liquids, fruits, and grains, were found on the same wreck and are also pictured in Figure 3.3. In commerce, these amphoras were as important as grinding stones.

Other methods were also used for breaking grains. A painting on the wall of a tomb in Egypt dating from about 1400 BC shows oxen treading husks from kernels and the husks being removed by wind currents in a winnowing process (see Figure 3.4). Because bread was the main food for rich and poor alike, it was all important in social and political affairs for more than 5,000 years. Caesar, Antony, and Claudius, for example, wanted Egypt for its grain (Jacobs 1944).

> *Bread reigned over the ancient world, no food before or after exerted such mastery over men. The Egyptians who invented it based their entire administrative system upon it. The Romans converted bread into a political factor, they ruled by it, conquered the entire world by it and lost the world again through it....*
>
> *It is hard to imagine that in the 2nd century AD Africa from Tunis to Tangier was one vast Roman field of wheat. The great accomplishment of the Romans consisted not in their introduction of Roman law or Roman police but in their making farmers of hundreds of thousands of nomads. (Jacobs 1944)*

Grinding was also used in antiquity to prepare pigments and drugs:

> *The Ebers papyrus, which was probably written about 1500 BC is the longest, most complete, and most famous of the medical papyri. ...About 700 drugs, made up into more than 800 formulas, are found in the Ebers papyrus. ...balances have been found along with mortars, mills and sieves used in preparing drugs. (Magner 1992)*

Mortars and Pestles

> *The mortar proper rested on a stand whose height was such that its upper rim reached about to the knee of the worker, or else it and the stand were made of one piece. The pestle was made of wood about 30 to 39 inches long and narrowed down at the middle so that it could be conveniently grasped. (Neuberger 1930)*

By 6000 BC, mortars and pestles were being used for size reduction. Rocks such as basalt or hard limestone were suitable materials for grinding grains, and very hard volcanic stones such as trachyte or sanidine feldspar were used for breaking ores. Impact and compression were the breakage mechanisms. Wear on stone mortars and pestles was always a problem, particularly with ores, and eventually the pestles were clad with or made entirely of iron. This reduced wear and allowed for heavier impacts that made the particles shatter faster.

The mortar and pestle is probably the only piece of equipment that has not changed in more than 10,000 years (see Figure 3.5). It is still an essential item in many laboratories and kitchens and is used to prepare food in developing countries. Over its long history, many methods have been used to operate the mortar and pestle to make the required effort as easy as possible. Figure 3.6 shows some methods of working mortars and pestles (Kuprits 1967).

Figure 3.7a shows an operator grinding gold ore in a mortar and pestle in 1999, a scene that differs little from the one depicted on a Turkish postcard (Figure 3.7b), which shows how ore was ground in several stages for metal extraction hundreds of years ago.

Saddlestone Mills

> *She spread some handfuls of grain upon an oblong slab of stone, slightly hollowed on its upper surface, and proceeded to crush them with a smaller stone. ...for an hour or more she laboured so, with arms shoulders and loins, in fact with her whole body till the operation was complete....*
>
> *The "saddle" corn stone stands pre-eminent among the milling appliances of the world. It derives its name from a resemblance its concave upper surface bears to the seat of a saddle. (Bennett and Elton 1898–1904)*

FIGURE 3.5 Mortar and pestle pictured on a Greek vase (Neuberger 1930)

FIGURE 3.6 Methods for working mortars and pestles in Russia (Kuprits 1967)

A B

FIGURE 3.7 (a) Grinding in Central Africa in 1999 (reprinted by permission from Malcolm Powell, University of Cape Town) (b) grinding in Turkey hundreds of years ago (reprinted by permission from MTA Museum)

Saddlestone mills seem to have first appeared in Chaldea (part of present-day Iraq), because "...no trace appears of their existing in any other country prior to their flourishing in Chaldea" (Bennett and Elton 1898–1904). These mills consisted of small rocks with smooth faces that were rubbed in a reciprocating motion on larger fixed rock slabs, either back and forth or at an angle when attached by one end to an upright. The upper stone did not roll. The fixed slabs were slightly hollowed and usually sloped slightly downward, away from the workers. The reciprocating motion was exhausting to maintain and wasteful of energy because particles that were already small enough were rebroken. Saddlestone mills, however, were better than mortars and pestles for producing very fine particles. Abrasion was the important breakage mechanism and was more effective than impact for producing very fine particles; the stones were always in contact, meaning that the particles were being worked on all the time.

With a saddlestone mill, one person could grind enough wheat to make bread for about eight people (de Camp 1970). This meant that for more than 2,000 years at least one-eighth of the world's population carried out the backbreaking work of grinding grain by hand. More labor was needed for grinding minerals because they were much harder. The lives of the grinders in households was hard enough, but the lives of slaves and prisoners who were required to grind grain to supply large organizations such as the palace and the army must have been short and appalling.

> *Good gods, what a sort of poor slaves were there; some had their skin bruised all over black and blue, some had their backs striped with lashes and were but covered rather than clothed with torn rags, some had their members hidden by a narrow cloth, all wore such ragged clouts that you might perceive through them all their naked bodies, some were marked and burned in the forehead with hot irons, some had their hair half clipped, some had shackles on their legs, ugly and ill favoured, some could scarce see, their eyes and faces so black and dim with smoke, their eyelids all cankered with the darkness of that reeking place, half blind and sprinkled black and white with dirty flour marks like boxers who fight together befouled with sand. (de Camp 1970)*

The Book of Judges (16:21; *Good News Bible* 1976) left no doubt that grinding was a severe punishment. In the story of Samson, the leader of the Hebrews was lulled to sleep by Delilah so that she could crop his hair after learning that his long hair was the source of his strength. In his weakened condition he was made a prisoner by the Philistines, his eyes were gouged out, and he was taken to Gaza bound with fetters of bronze and "...set to grinding corn in the prison." Little wonder that when his hair grew and his strength returned, he wreaked great vengeance on the Philistines by grasping two columns of the temple and fracturing them, killing many who had gathered to make sport of him.

Medieval painters used saddlestone mills to produce pigments, and Cennini described the technique (Research Association of the British Paint, Colour and Varnish Manufacturers 1953):

> *To grind properly, procure a slab of porphyry which is strong and firm. ...Take another stone, also of porphyry, smooth on one side, and raised on the other, in the shape of a porringer and half the height of one, of such a form that the hand may hold and guide it at pleasure. Then take some of the colour and put it on the slab, and with that stone which you hold in your hand break the pigment into small pieces. Put some clean water either from a river, a fountain, or a well, to the colour and grind it well for half an hour, or an hour, or as long as you please: but know that if you were to grind it for a year, so much the better would be the colour.*

FIGURE 3.8 Saddlestone mills ([a] Bennett and Elton 1898–1904 [b] Perrogon 1994)

Saddlestone mills (shown in Figure 3.8) were the main fine-grinding devices for more than 1,500 years. Their importance diminished after the quern was invented, but they were still being employed in a few places in the 20th century.

Rotary Querns

Mortars and pestles and saddlestone mills had several problems: the exhausting motions required of the operator, the batch nature of the processes, and the unnecessary rebreaking of particles that were already small enough. The invention of the rotary quern, probably around 600 BC (estimates of the date of its appearance vary by several hundred years), solved these problems. Querns, which used shearing and compression forces, had a fixed lower stone and an upper stone that rotated on a pin, so that the moving and fixed surfaces were just clear of each other. The hole in the lower stone of the early quern contained a spigot on which the upper stone spun, just clear of the lower stone.

> *Originally the upper stone was probably always lifted off when fresh grain was added. Later however the upper stone had a hole in the centre while the lower had a projection which passed through this hole. Between this pivot and the hole enough space remained for adding grain when necessary. In order to spread the grains entering at the central hole over the whole space between the two millstones the latter had small grooves running out radially between which further grooves were inserted inclined at acute angles to them. (Neuberger 1930)*

Compared with saddlestone mills and mortars and pestles, rotary querns had several advantages. They could be operated manually in a rotary motion by a crank, which is a vertical shaft attached to a horizontal spoke. The crank, which was an important engineering advance itself, enabled the stone to be turned more easily:

> *Crank motion first appeared in querns among people who still ground their grain at home. The quern maker drilled a vertical hole near the edge of the upper stone and stuck a peg in the hole. The operator could then spin the upper stone round and round more easily than she could with radial spokes where she had to change her grip with every revolution. In fact she could turn out about a bushel of flour every day–twelve times as much as before. (de Camp 1970)*

Figure 3.9 shows how querns evolved. At first they were small hand-operated units in family dwellings that made enough flour for a single family. As they came to be used by several families and then by whole villages and towns, ever-larger machines with higher production rates were built. The mills at Pompeii, shown in Figure 3.10, were 2 m high, and these mills, and the ovens that were built beside them, were the core of a large

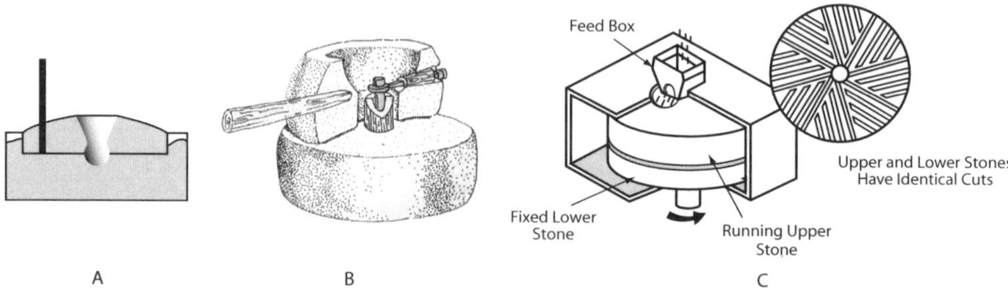

FIGURE 3.9 (a) Early batch quern powered by a crank (Bennett and Elton 1898–1904) (b) continuous quern (Australian Wheat Board 1998; reprinted by permission from AWB Ltd.) (c) pattern of grooves cut into a millstone (Bass 1988; reprinted by permission from American Association of Cereal Chemists)

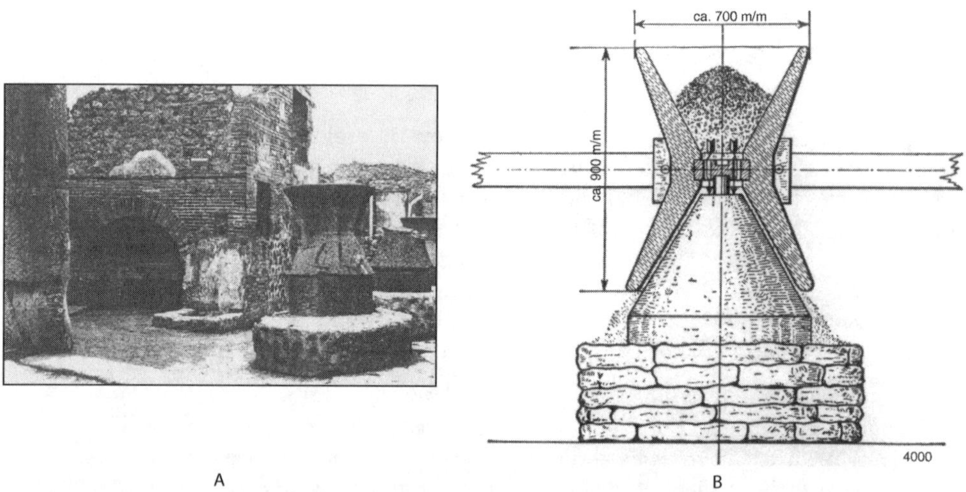

FIGURE 3.10 (a) Bakery at Pompeii showing the remains of the mills (Ward-Perkins and Claridge 1980) (b) typical mill (de Camp 1970)

commercial milling and baking complex. In addition to the higher production rates, the quality of the flour was higher (Ward-Perkins and Claridge 1980).

> *Hellenistic and Roman mills finally took the form of the hour-glass mill. While the lower stone was conical the upper was cylindrical. This cylindrical stone was hollowed out to fit the cone below and had a similar cone-shaped hollow above, into which the grain was fed. An adjustable wooden pivot made it possible to regulate the clearance between the stones and thus to grind the grain to the desired fineness. The hour-glass mills at Pompeii were huge affairs 2 m (6.5 ft) high weighing hundreds of pounds. Capstan spokes projecting horizontally made it possible to turn the upper stone. (de Camp 1970)*

In the mills at Pompeii, men turned the upper millstone in the hourglass type of mill by pushing on spokes that were inserted in slots. Boring holes in tough stones was not easy with the tools of the day. Beedell (1975) suggested that people bored holes by spinning a pointed hardwood stick with a bowstring and pouring wet sand around it as an abrasive. The pilot hole could then be expanded.

FIGURE 3.11 Querns driven by animals (de Camp 1961)

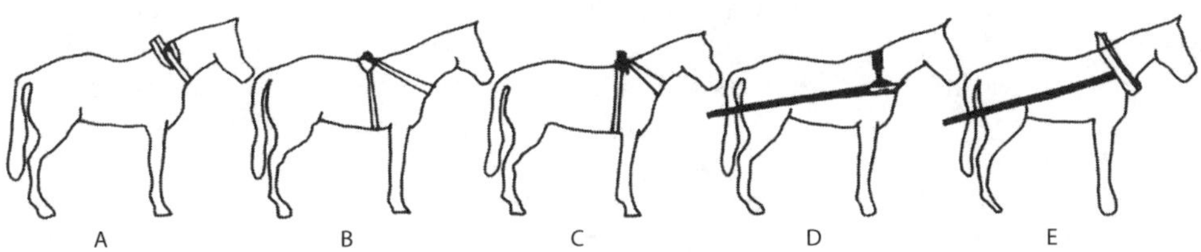

FIGURE 3.12 Evolution of harnesses for making use of the power of animals: (a) neck yoke—pulling from shoulders (b) dorsal yoke—pulling from chest (c) traction harness—pulling from lower part of neck (d) shoulder-collar harness (e) breast-collar harness (Clutton-Brook 1952; reprinted by permission from British Museum)

In time, animals came to be used as a source of power. Although animals cannot drive a machine that has a reciprocating motion, they can be harnessed to a shaft and drive a rotating stone by walking in a circle. Figure 3.11 shows one method of using animals to drive a quern. By 200 BC mules and donkeys were replacing men for this task, although the use of animals was relatively uncommon because slaves continued to be the cheapest and most flexible source of energy.

A difficulty with using animals was in designing a harness that would allow the animal to effectively transmit its power to the machine. The first harnesses were placed around the neck of the horse, but as loads increased, these harnesses tended to tighten and even strangle the animal. Gradually the design changed to relieve the stress on the neck, eventually reaching the point at which full power could be delivered through the modern shoulder-collar and breast-collar harnesses. Figure 3.12 shows the evolution of the harness (Clutton-Brook 1952). Although the use of animals to power grinding querns had limitations, it was an important development, because it was humankind's first extension of the use of animal power since animals were harnessed to plows and carts 3,000 years earlier.

The skill of the craftsman in stone working was a crucial factor in the evolution of the quern. These skills can be admired in the magnificent stonework in the pyramids and monuments in Central America and in the eastern Mediterranean, but they were also necessary for building the great querns in cities such as Pompeii. The high production rates of these querns led to the commercial baking industry. By the latter part of the 4th century, there were 254 bakeries in Rome (Mosse 1969).

Querns were used for grinding grains until late in the 19th century, when rolls were found to be more efficient for producing high-quality flour. Rolls now dominate grinding for the production of flour, but water-driven querns can still be found in many communities. Two types of crushers that are widely used today had their origins in querns: (1) disk crushers, which are mainly used for corn, and still feature the grooves from earlier times; and (2) cone crushers, which are used for ores and aggregates.

The quern was an outstanding invention. It greatly increased the output of workers and "constituted the earliest complete grinding machine: the first in which the parts were combined into a perfect mechanism" (Bennett and Elton 1898–1904). It made the grinding of wheat so much easier that it became essential in every household. "The two ever present sounds of households were the clack of the loom and the grate of the hand mill" (de Camp 1970).

Limitations of Querns Querns were heavy and expensive, because they were made of large, tough stones weighing up to hundreds of kilograms each. With care they would last for many years, even generations. Suitable stones could be found in a limited number of locations, and a significant trade developed in stones for querns and for other grinding devices such as hopper-rubbers. Considerable skill was needed to set the quern stones so the machines would break the particles efficiently with minimum wear. If the gap was too large, the corn was not ground properly; if too small, the corn was burnt; and if the stones were set to touch, they wore out quickly.

And there was always the problem of power. Grinding was carried out under duress by prisoners and slaves, only occasionally by free men or women with no other means of livelihood. Grinders generally had a short life expectancy because of the harshness of the task, the grim living conditions, and the unending work in a dusty atmosphere. New slaves were always needed to grind raw materials for flour and metals. When the Roman Empire was expanding through conquest, this was less of a problem because plenty of captives could be forced into slavery. It has been estimated that by 28 BC there were 40 million free citizens and 20 million slaves in Italy and as many more again in other countries of the Roman Empire (Hopkins 1978). A population of this size required considerable food to keep it functioning, and many slaves were needed to grind the wheat to produce the bread.

But when the Roman Empire expanded to include the principal nations of Europe, Africa, and West Asia, and the people from those areas had been enslaved, the sources for slaves dwindled and there were no more countries to be conquered. Other methods had to be devised to supply an adequate source of energy for grinding, which led to the harnessing of water as an alternative energy source.

MUSCLE POWER FOR MINING

Querns for Grinding Ores

Ore breakage in Greek and Roman times was mainly by human muscle power, although cattle mills were introduced about 200 BC. The usual procedure was for rocks to be broken with hammers, large pebbles to be broken in mortars and pestles up to 1.5 m deep and 4 m in diameter, and small pebbles to be broken in rotary querns. Screens made of hazel twigs or horsehair were used to separate the fine particles.

The Japanese woodcut in Figure 3.13a shows an example of how the rotary quern was used to break ores, with more detail illustrated in Figure 3.13b. The ore grinders took ore that had been broken to 5-mm pieces in the mortars and broke it further to

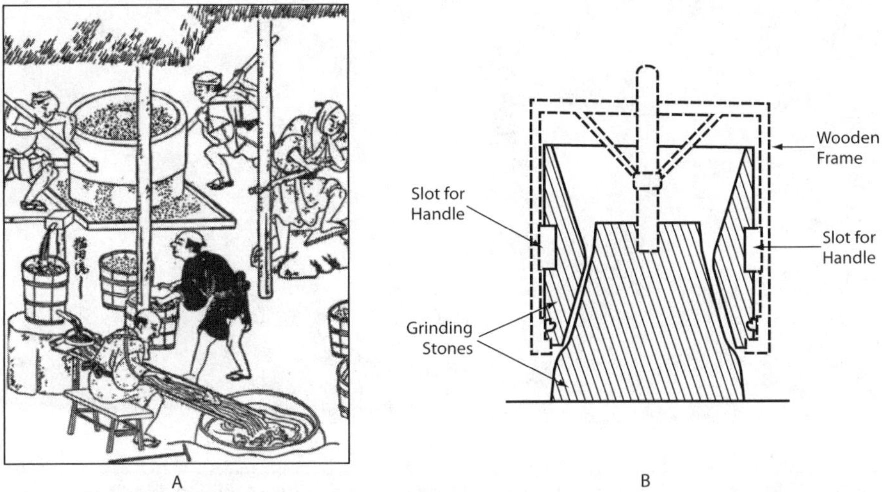

FIGURE 3.13 Querns driven by men: (a) Japanese woodcut that shows ore being ground in a quern and passing down a launder (Lebeter 1949a; reprinted by permission from *Mining Journal* [formerly *Mine & Quarry Engineering*]) (b) construction of a manual quern (Healy 1978; reprinted by permission from Thames and Hudson)

0.25-mm pieces in the rotary querns. A mill produced about 4 tons per 24 hours, was driven by up to six slaves, and used about 0.5 kW of energy. The coarse product from the screen was returned to the mill for further grinding.

Men and women grinding ores must have worked in conditions that were even worse than for those grinding grains because the pebbles were much harder. The conditions were described in a footnote in *De Re Metallica.*

> *Those over 30 years of age take a piece of rock of a certain quantity, and pound it in a stone mortar with iron pestles till it be as small as a vetch; then those little stones so pounded are taken from them by women and older men, who cast them into mills, that stand together there near at hand in a long row, and two or three of them being employed at one mill they grind a certain measure given to them at a time until it is as small as fine meal. No care is taken of the bodies of these poor creatures, so that they have not a rag to cover their nakedness, and no man that sees them can choose but commiserate their sad and deplorable condition. For though they are sick, maimed or lame no rest or inter-mission in the least is allowed them; neither the weakness of old age, nor women's infirmities were any plea to excuse them; all are driven to their work with blows and cudgelling, till at length, overborne with the intolerable weight of their misery, they drop down dead in the midst of their insufferable labours. (Agricola 1950)*

The multiholed stone was used 3,000 years ago to grind ore that had already been broken to a powder of less than 5 mm for mineral separation. High-grade particles were placed in a depression in the large fixed stone and broken with a smaller rounded stone held in the hand. When the depression deepened to the point at which breakage was difficult, the operator moved to a new position on the stone.

Hammers and Gads and Fire Setting

In early human history, rocks used for buildings, monuments, or metal extraction were mined by hand using hammers, levers, and wedges made from stone, wood, or bone.

A B

FIGURE 3.14 Hammer and gad mining: (a) 15th-century woodcut (Agricola 1950) (b) 18th-century drawing by Eduard Heuchler showing mining and manual breakage. One exhausted miner is resting (Paul 1970).

This was the first stage of size reduction, and the procedures used are shown in the woodcut and drawing in Figure 3.14. They varied little from the Stone Age to the 17th century when explosives were first used to break rocks. The hammers and gads used 100 centuries ago are still in use today, albeit with good-quality steels instead of stone. Like the mortar and pestle, they will always be used.

Fire was introduced at least 4,000 years ago to crack the rock and make boulders easier to remove (Drier and Du Temple 1961). Figure 3.15 shows fire setting as it was practiced in 1500 AD. Wood was burned against the rock face for hours until it became very hot. It was then cracked naturally or water was thrown against the rock to induce cracking. Pebbles and boulders could then be easily pried away. The intense heat and toxic fumes made fire setting underground a dreadful task, but it could be controlled and was safer than explosives. It was still used in tunneling and underground mining in Europe and Japan late in the 19th century with advances of 1.5 m to 6 m per month being achieved. Fire setting became legendary with Livy's story of Hannibal using fire and vinegar to shatter rocks in the Alps to make a track for his elephants (Drinker 1888).

Manual Crushing

The second stage of size reduction in mining was manual crushing of mined rocks to sizes suitable for direct use or further breakage (see Figure 3.16). Manual crushing was widely used until the end of the 19th century and is still used in many regions of the world. By 1900, the nomenclature for hand crushing was well defined (Foster 1894):

- Ragging: breaking very big rocks with large sledgehammers weighing 3.5–5.5 kg
- Spalling: similar to ragging but breaking small rocks with 1.8- to 2.3-kg hammers
- Cobbing: using a small hammer to knock waste from rocks
- Bucking: using a broad, flat hammer, about 100 mm square with a flat face, to reduce ore to coarse powder

Men did the ragging, but women and even children were involved in spalling, cobbing, and bucking. The drawings in Figure 3.16 put these exhausting occupations in a much more elegant light than existed in practice. The same may be said of "bal maidens," the name by which the women who carried out the cobbing in the tin mines of

FIGURE 3.15 Woodcut of fire setting as practiced in metal mines in Saxony in 1500 AD (Drinker 1888)

FIGURE 3.16 Forms of hand breakage: ragging, spalling, cobbing, and bucking (Foster 1894)

Cornwall were known. It may have been a charming name, but there was nothing charming about the work.

Breaking rocks by hand was always a difficult and disliked occupation, but it was necessary in both civil and mining engineering. In England,

> *...a certain quantity of broken stone was frequently exacted as a task from vagrants in casual wards as a return for food and lodging; the work was, however, so greatly hated that those wards that imposed it were studiously avoided by the fraternity. (Dickinson 1945)*

Yet, there were a few who regarded hand breakage in a more kindly light. "The Roadmender," for example, was a wonderful, if unintentional, tribute to the men who broke stones on the roads (Fairless, circa 1900). The author of this essay eloquently described the observations of a philosophical roadmender who worked alone and took advantage of the silence and loneliness to observe the passing parade as he broke and laid the stones, "...noting the changes that progress brings even in that quiet countryside." She finished her work with this observation of the roadmender: "I have learnt to understand dimly the truths of three great paradoxes–the blessing of a curse, the voice of silence, the companionship of solitude." This was a gracious comment about an activity that brought untold misery to countless numbers of people.

Manual Drilling

Breaking rock by blasting with explosives started in the 17th century and slowly became part of mining practice. This required that holes be bored to put the explosive inside the rock. Rock drilling was not a new art: "The use of diamonds for rock boring was known to the ancients" (Drinker 1888), and "The ancient Egyptians are said to have used corundum dust or pebbles in order to drill holes in porphyry" (McGregor 1967). The drill holes in ancient times would have had small diameters but holes for explosives had to be larger. Until 1696, holes were 50–62 mm in diameter, 1 m deep, and held 1 kg of powder (Drinker 1888).

It was estimated that with good hammers and bits three men could drill a hole 2-m deep in granite in 5 to 6 hours depending on the rock and the location where the work was done.

> *In single hand hammer drilling the bit was held and rotated in one hand struck by a 1.8 kg hammer held in the other for drilling 32 mm size hole. For drilling deep holes a double hand drill was essential. One man held the bit, the second man struck it by a 4 kg hammer. (Chugh 1985)*

CONCLUSION

Since the dawn of human history, we have needed to efficiently reduce the size of materials to make usable products. The pressures driving this technology's development–the need for larger, more reliable, and more efficient sources of energy; for larger, more efficient machines; for wear surfaces with longer lives; and for machines and methods that produce finer products–have always been the same. Only the techniques have constantly changed and improved to meet the challenges of all the differing, and sometimes competing, technology requirements.

CHAPTER 4

Water Wheels, Windmills, and Beyond

GRINDING GRAINS INTO FLOUR

For thousands of years, grinding grains to make flour was the primary application of grinding technology. Beginning in about 100 BC, the development of mills driven by water and wind revolutionized the practice of agriculture. Although the importance of bread is underrated today, it is still the main food in many countries (Table 4.1).

WATERPOWER FOR GRINDING

The Water Wheel

The water wheel was one of the great engineering inventions, using energy from renewable natural resources for the first time. Harnessing waterpower caused a peaceful revolution that changed society irreversibly.

Water wheels were first used to drive rotary querns. Legend has it that this occurred in the eastern Mediterranean about 100 BC, a story that may have its roots in an epigram written by Antipater of Thessalonika around 85 BC, which recorded the significance of the miracle of waterpower. Antipater's ode to the water wheel and the grinding mill (Bennett and Elton 1898–1904) is all the more pleasing because engineering achievements are rarely recognized in poetry:

Ye maids who toiled so faithful at the mill,
Now cease your work and from those toils be still:
Sleep now till dawn, and let the birds with glee
Sing to the muddy morn on bush and tree:
For what your hands performed so long and true,
Ceres has charged the water-nymphs to do.

Pompeii's soldiers found water wheels in the palace of King Mithridates VI after they conquered Pontus in 63 BC. It is not surprising that they were used there, because the kingdom was in a mountainous area on the southern coast of the Black Sea where many fast-flowing streams made it a suitable place for water wheels. But the Romans did not use waterpower widely until the shortage of slaves made it essential about 400 AD, and more than 1,000 years were to elapse before it became a general source of industrial energy.

TABLE 4.1 Consumption of bread in various countries in 1988

Consumption Range, kg/head/yr	Countries
150–200	Libya, Bulgaria, Egypt, Algeria, Morocco, Roman Empire (200 BC–400 AD)
100–150	Soviet Union, Greece, Hungary, Poland, Czechoslovakia
70–100	East Germany, Portugal, Belgium, Italy, Spain, West Germany
50–70	Ireland, France, Netherlands, New Zealand, Austria
20–50	Great Britain, Australia, United States, Sweden, Switzerland, Denmark

Source: Modified from Orth and Shellenberger 1988.

Water wheels can be operated with the wheel rotating in a horizontal plane driving a vertical shaft or in a vertical plane driving a horizontal shaft (Figure 4.1). The water wheel Antipater saw probably had a vertical shaft, because gears would not have been available and the shaft could be attached directly to the rotating stone. This type of mill was called a Greek water mill, no doubt because it was invented in a province of Greece. Mills driven by vertical wheels were called Roman water mills, probably because of their use by Roman engineers. Vertical wheels were more efficient than horizontal wheels because they could be built larger, and larger mills draw more power and grind more materials. In addition, the gearing allowed mill speeds to be changed. Vertical water wheels became the leading source of industrial power, dominating for about 1,500 years until the rise of steam power. Eventually, waterpower allowed inventors to design machines that greatly increased the volumes of metals, food, and fabrics produced. With water as the power source, engineers could undertake tasks that had previously been impossible, such as the large-scale production of iron and the dewatering of deep mines.

If there was a single key element distinguishing Western European technology from the technologies of Islam, Byzantium, India, or even China after about 1200, it was the West's extensive commitment to and use of waterpower. For centuries the vertical water wheel was of critical importance, not only to western technology and the level of western industrial output, but also to the way western society evolved (Reynolds 1983).

Today, water wheels are little used as direct sources of power, although water is extensively used to drive hydraulic turbines (which are essentially large water wheels) for generating electrical power. But water mills with either horizontal or vertical wheels are still used for grinding grains in many countries.

Horizontal Water Wheels The most important requirement for a Greek mill was a fast-flowing stream of water. There were many suitable streams in the mountainous regions of Greece and Turkey, and in Asia as far afield as India and China. So it is not surprising that mills similar to the ancient Greek mills have been found far and wide, but we do not know whether they came from one source or emerged independently in different locations. A technique commonly used to drive water mills, which are querns driven by water wheels, is shown in Figure 4.1. Water is delivered through a pipe to the blades of a horizontal water wheel, and the wheel rotates a vertical shaft that drives the rotating stone of a quern.

Greek water mills are still the cheapest method of grinding grains in many villages. A mill driven by a horizontal water wheel near Corum in central Turkey has grinding stones that are 1.12 m in diameter and 0.26-m thick with a 0.22-m hole in the center. The stones are 20 years old and are used 12 hours per day for 5 months a year. The owner expects the stones to last for another 20 years. The mill produces either animal feed at 170 kg/hr or flour for human consumption at 100 kg/hr, and the owner is paid one-eighth of the value of the product.

The Greek water mill, like the mortar and pestle, is one of the few devices in use today that would be recognized instantly by those who operated them more than 1,500 years ago.

Vertical Water Wheels Vertical water wheels could take three forms (shown in Figure 4.2):

- The undershot wheel, in which the water struck the paddles on the wheel at its base and moved them at the same velocity as the water
- The overshot wheel, in which water filled the buckets on the wheel at the top and moved them by the velocity of the water and the potential energy of the filled buckets
- The breastshot wheel, in which water filled the buckets about halfway up the wheel and moved them by the potential energy of the filled buckets

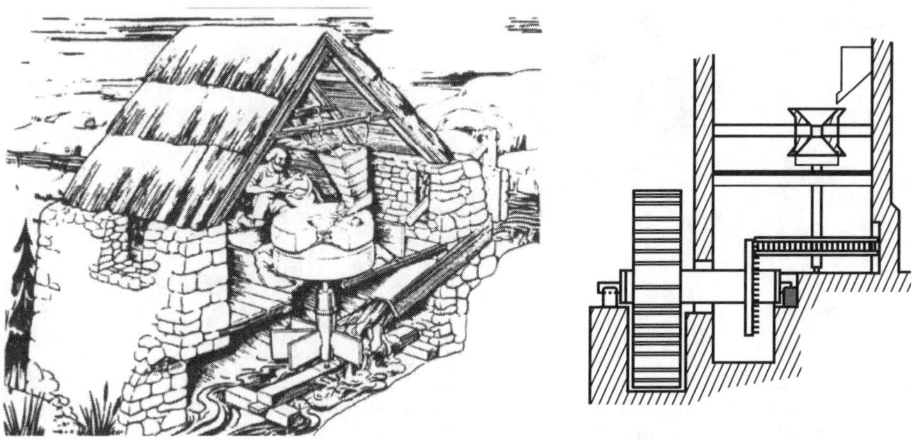

FIGURE 4.1 (a) Horizontal (Greek) and (b) vertical (Roman) water mills ([a] Reynolds 1983; reprinted by permission from University Minnesota Press [b] Reynolds 1983; reprinted by permission from Miriam Usher Chrisman)

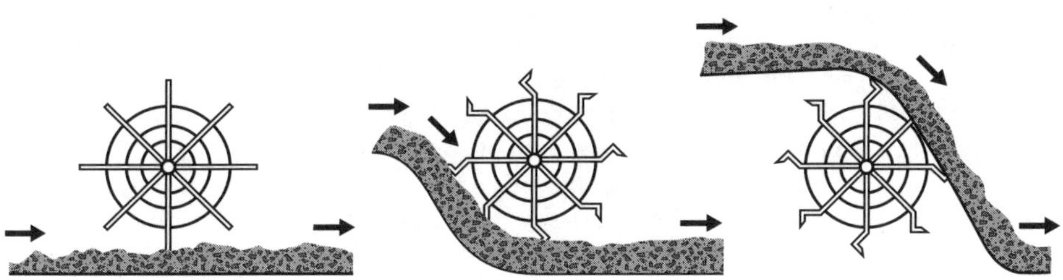

FIGURE 4.2 (a) Undershot or Roman vertical water wheel (b) the breastshot water wheel and (c) the overshot water wheel (O'Brien 1965)

The undershot vertical water wheel described by the Roman engineer Marcus Vitruvius Pollio in 25 BC (Pollio 1931–1934) was probably invented about 50 years after the Greek mill.

The overshot wheel typically achieved efficiencies of 50% to 70%, much more than the undershot wheel at 15% to 30%, but it was more expensive because it required a dam, millrace, sluice gates, and a tailrace (Gille 1966). Despite their higher efficiency, water mills were not used much for grinding for a few hundred years after the vertical water wheel was developed—probably because slaves were cheap and plentiful, could operate querns efficiently by hand, and could fix problems that arose. But slavery slowly declined in the Roman Empire, and the Emperor Constantine, who ruled from 324 to 337, eventually issued an order that freed the slaves. This order was often evaded until an imperial edict was issued in 398 AD, which stringently enforced the protection of slaves, and then the need for another source of energy became urgent.

So it was that the Roman Empire gradually moved toward waterpower for grain grinding. Small water wheels were easy to build, but their capacity was limited. Large water wheels were expensive, but they could generate considerable power and drive high-capacity grain mills. Consequently, by the 4th century, Roman engineers were

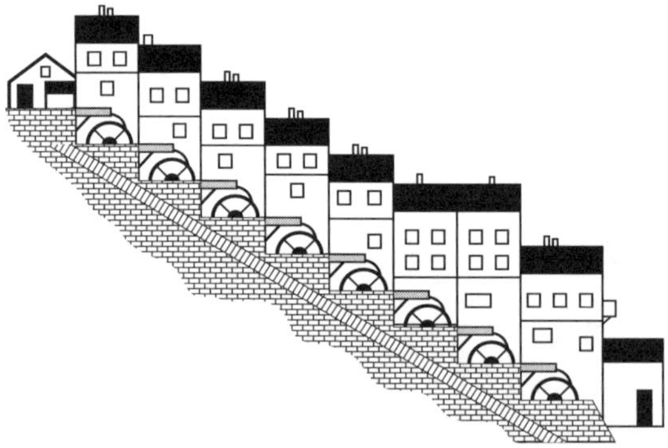

FIGURE 4.3 Fourth century flour mill at Arles, France, which used eight pairs of 2-m (7-ft) diameter water wheels (O'Brien 1965)

building large water mills. One grain mill in Arles, France, had eight pairs of overshot water wheels (2 m in diameter) that operated in series, generating about 3 kW per wheel and producing flour for 80,000 people (Reynolds 1983; see Figure 4.3). The fall of water was 19 m, and each of the 16 wheels ground 2.5–3.0 tons in 10 hours. The mill at Arles did the work of about 1,500 slaves (Gimpel 1976).

Water mills were used extensively in the Roman Empire by the 4th century and were still being used in the 19th century (see Figure 4.4).

The mills in Rome were located in Janiculum, which is a hilly region on the western side of the Tiber River. An aqueduct brought water to the crest of the hill, and, as it flowed down the incline, gained sufficient velocity to drive the mills. During their siege of the city in 537, the Goths cut the water supply to starve the defenders, but the Byzantine general Belisarius built a floating undershot water wheel on the Tiber River to drive a flour mill that was placed on a boat. The invaders were repelled, at least for the time being (Reynolds 1983).

The vertical water wheel had two important effects on engineering: It increased the capacity of processes because it made more power available, and it demonstrated the importance of gears in drive mechanisms. It is not known when the gear that transmitted power from the vertical water wheel to the millstone was invented, but its use with querns was a significant engineering advance in its own right. The vertical wheel and the gear set the scene for the invention, 1,500 years later, of the water-powered stamp mill, which was extensively used for 400 years and was one of the longest-lasting size-reduction machines ever invented. After the collapse of the western Roman Empire in the 5th century, during the 500-year period known as the Dark Ages, waterpower spread throughout Europe; the type of flour mill used in medieval times is shown in Figure 4.5.

Little progress was made in the arts and sciences during the Dark Ages because the people who controlled the remnants of the Roman Empire had other interests. But grain mills were important, because there was no alternative to flour for bread making. We can surmise that grain milling improved during this 600–1,100-year period although few records exist about its progress. Water wheels did continue to proliferate as indicated in the records of the Domesday survey (The Domesday Book Online 1999–2004), which shows that, by 1086 in England, 5,624 water mills existed in 3,000 locations. Most were used to grind grain for about 1.4 million people (Gimpel 1976).

> *Because building a flour mill was quite expensive, a landlord, the church, or royalty usually owned the mills. Each farmer who brought grain to the mill paid a fee*

FIGURE 4.4 Floating mill in Armenia in the 19th century with the undershot water wheel (Bennett and Elton 1898–1904)

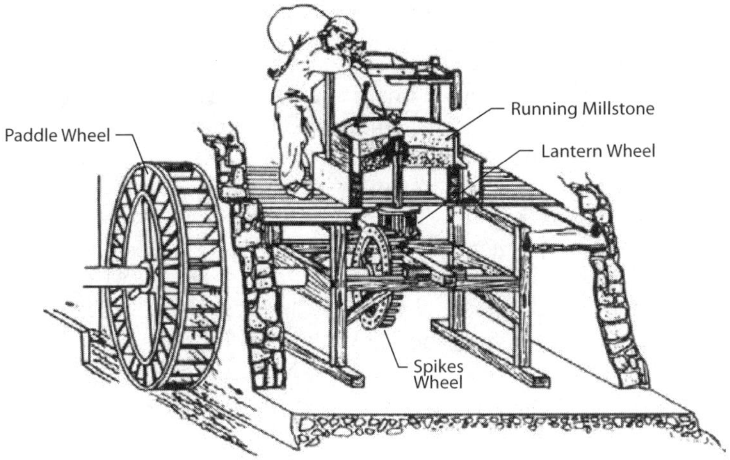

FIGURE 4.5 Illustration of a flour mill used in medieval times (Godn and Wilm 1994)

or toll. In Europe, many of the smaller querns were destroyed in an effort to force people to use the king's or church's official mill. This also forced families to buy flour rather than being able to produce a small quantity for their own use (Planet Wheat 2002a, 2002b).

Grain mills driven by vertical water wheels still operate commercially in many parts of the world, although they are principally rehabilitated mills. Their revenue comes from sales of stone-ground flour and from fees paid by tourists who enjoy visiting the mills and watching the ancient grinding process. The Josephine mill in Cape Town, South Africa, near the famed Newlands cricket and football grounds, is typical of mills of this type. It has a breastshot water wheel that is 7.9 m in diameter and 1.4 m wide, with 70 buckets. On start-up, the wheel takes 12 L for the first three buckets and then operates

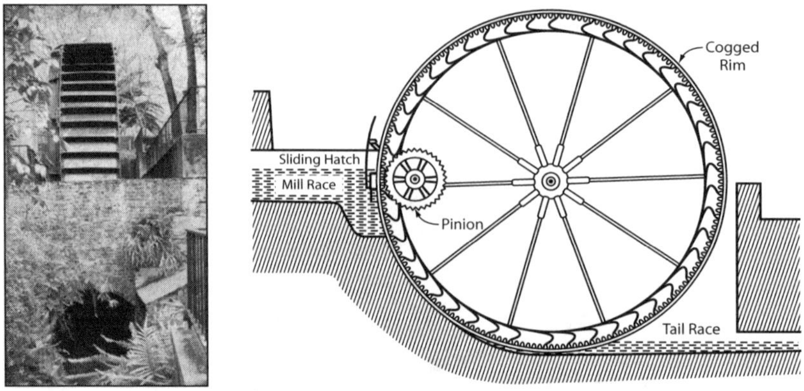

FIGURE 4.6 Water wheel at the Josephine mill in Cape Town, South Africa, showing the buckets and the method of transmitting power from the wheel to the driving shaft (Walton 1978)

on 1 L per bucket. The grinding stones are 1.6 m in diameter. When the mill was in full operation, a pair of stones produced about 2 tons of flour in 24 hours (Figure 4.6; Walton 1978).

WIND POWER FOR GRINDING

Historically, the wind has also been an important source of renewable energy for driving grinding mills. "Crude vertical-axis panemones have ground grain in the Afghan highlands since the seventh century" (Gipe 1995). Panemones had walls to direct the wind past vertical sails that were attached to a rotating axle, as shown in Figure 4.7.

Windmills existed in arid areas of Persia in the 10th century, were used in France by 1180, and came into use in England by 1185. They became a significant source of power in locations where the wind was fairly dependable. Gipe (1995) reports that, by the 19th century, 500,000 windmills were used in China, with an equal number in use in Europe. Wind power was used for general grinding, including: corn and grains; flint for making china, gunpowder, and clinker; chalk for making whiting; and minerals for dyes and paint. It was also used for many other duties such as pumping, sawing, and pressing oils from seeds. Figure 4.8 shows a windmill and the mechanism for driving the querns. Early photographs of the Netherlands feature large windmills being used to pump water from land below sea level back into the sea.

> *The windmill, as perfected by the British millwrights of the 19th century, was an extraordinarily sensitive and sophisticated machine. The vagaries of the wind, always a fickle and uncertain source of power, made windmills more complicated and less reliable than the water mill and demanded very different skills of the miller. In the water-powered corn mill it was possible, by adjustment of the hatches, to achieve a more or less uniform speed and thus ensure an even texture in the meal produced. The windmill, however, demanded a different solution. The force of the wind was constantly changing, and this had a curious effect on the action of the stones. As the speed of the mill increased, the runner stone would tend to rise on its spindle, opening the gap between the grinding surfaces and making constant adjustments necessary in a choppy breeze (Reynolds 1970).*

Grinding grain in stones driven by the wind must have seemed like a dream come true. The mill rotated to the music of the wind in the sails, and the grinders worked under

FIGURE 4.7 Panemone—an early windmill (Gipe 1995)

FIGURE 4.8 Windmill and the internal mechanism showing the grinding stones (Reynolds 1970)

conditions far removed from the terrible conditions many had suffered for thousands of years. John Oliver of Sussex was one miller who appreciated the new technology. An excerpt from *Corn Milling* evoked the idyllic lifestyle that he seemed to enjoy for much of his career:

> *The vicinity of a mill may have often been desired as a place for burial: but so far as we know there was but one miller who, entertaining such a desire, had it realized–John Oliver of Highdown Hill, Sussex, who, in 1766, he being then 57 years of age,*

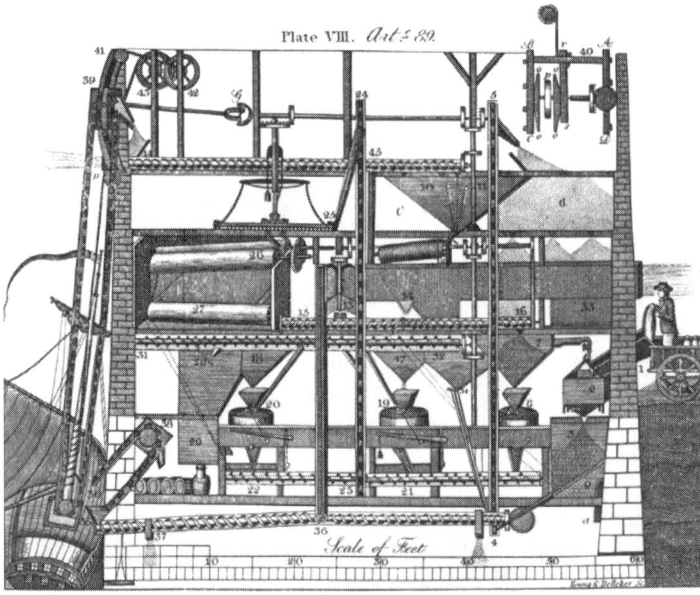

FIGURE 4.9 **Automated flour mill that Oliver Evans built in the United States in 1785 (Reynolds 1970)**

FIGURE 4.10 **Typical flour mill in United States in the 1870s which used querns driven by steam engines (Peterson 1976; reprinted by permission from Milwaukee County Historical Society)**

prepared his tomb beside his mill on the hill, and erected a summer house beside it, where in quiet hours he sat in meditation. For seven and twenty years he thus alternately tended the mill and the grave, till, in 1793, when he ceased his labours, he was obediently carried from the one to the other. ...He left by will 20 pounds a year for the tending of the tomb and the summer house, but of their ultimate fate we have no knowledge. (Bennett and Elton 1898–1904)

Millstones are still used in water-powered flour mills in villages close to reliable water sources and for making stone-ground, whole-grain flours in rehabilitated grinding mills using stones to make the grind.

THE AUTOMATIC FLOUR MILL

A seminal change in flour milling occurred in 1785 when Oliver Evans built an automatic flour mill to eliminate manual labor in material handling (Figure 4.9). Previously, several stages of grinding had been required, which included manual sifting between the stages to remove finished flour and inedible particles and to winnow out coarse materials for further grinding. In the automated mill, however, men and buckets were replaced by elevators to take the grain to the top of a tall building, and by downpipes and conveyors to move partially processed grain from machine to machine as it flowed from the top to the bottom of the building.

The main items of equipment in the automated mill were the

- Elevators, conveyors, and chutes by which grain and its products were transported
- Cylindrical reel separators that Evans invented to make sieving continuous
- Grinding stones
- The "hopper boy," which cooled the ground meal in a hygienic manner

The automated hopper boy was named after the human hopper boy, whose job in the pre-1785 plants was to rake the freshly ground flour manually to cool it before further processing. If not done quickly and thoroughly, the flour could be contaminated by insects (McAndrew 2000). Evans' hopper boy was a drum containing a slowly rotating rake into which hot flour flowed from the elevator and from which cool flour flowed to the sifter. Evans changed flour milling from a labor-intensive process that was unhygienic, with semiprocessed flour waiting in open piles for manual transportation to the next station, to a continuous, hygienic, and automated process.

STEAM POWER AND THE FIRST ROLLER MILLS

The advent of steam power greatly improved the efficiency of flour mills. In 1786, two 37-kW steam engines were connected to drive 20 stone mills (1.5 m in diameter) at the Albion grain mill in London (Fischer 1944), and for the next 100 years, grain was ground mainly in steam-driven querns. A flour mill that used steam-driven querns in the United States in the 1870s is shown in Figure 4.10. Soon afterward, roller mills began to replace the querns.

Both the querns and roller millers were found to be well suited for use in the "gradual reduction process" for making flour (see sidebar), which came to the fore in Hungary in the years from 1840 to 1870. At the time, rapid progress was being made in process engineering in nearby Austria and Germany, including advances in grinding technologies for the cement and ore industries. The intellectual environment of the industrial community has always been a definitive factor in innovation and progress, and industry was particularly progressive in central Europe during that time period. The Hungarian system was

THE GRADUAL REDUCTION PROCESS

Flour milling involves breaking the wheat grain to separate the endosperm from the bran (see Figure 4.11) and grinding the endosperm to flour. The problem is that much of the bran is not easily accessible because it is in the section of the grain that contains the germ, and cutting and peeling cannot remove the bran. Instead, fine grinding is necessary to liberate the bran for removal. The ancient methods of lightly breaking grains in a quern, sieving to free the bran, and grinding the remaining bran and endosperm again produced a fine flour that was edible but discolored and unattractive. Even though whiter flour could be made, more grinding and sieving was required, which drove up the price. Flour milling began to change in response to the growing demand for white flour in more prosperous countries.

By 1820, the continental wars had temporarily ended in Europe, and flour millers were able to invest time and money to improve the technology.

> *French engineers have introduced many original inventions, chiefly in the sphere of transportation, cleaning of grain, and dressing of the product. ... One of the greatest inventions of the French of that time is the cleaner and separator, the most indispensable machine of the grain cleaning department. (Kozmin 1917)*

Silk cloth proved to be more efficient for sieving than other materials, and millers used it to improve flour milling by operating querns in series with the first stones set well apart and the second stones closer together. The bran was captured from the ground products by sieving with silk—a system started late in the 18th century by the French miller Pigeaud. The combination of grinding in stages and using sieves to separate the ground products into size fractions became the basis for Hungary's gradual reduction process. The prospect of producing large quantities of white flour came into view, and the subsequent inventions of purifiers and steel roller mills brought the picture into focus.

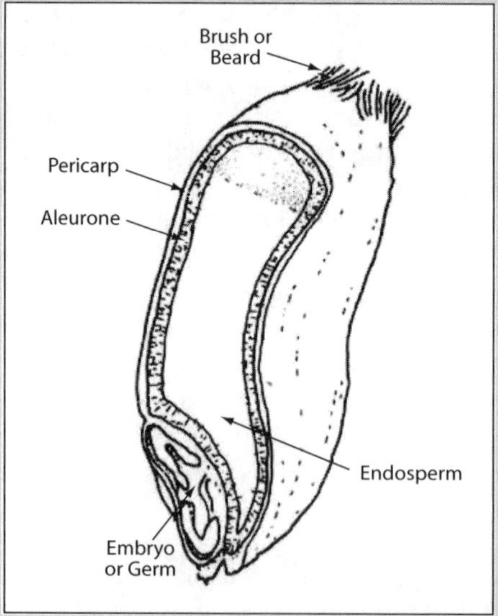

FIGURE 4.11 Schematic view of a wheat grain

introduced in Minneapolis, Minnesota, in the early 1870s. In the years that followed, the gradual reduction process was quickly accepted by new flour-production facilities in the United States, where mechanical development of the roller mills continued.

W.D. Gray, superintendent of the milling machinery department of E.P. Allis & Company (the predecessor of Allis-Chalmers), built and installed roller mills in a flour-production plant in 1878. His mills were so good that they changed little for 50 years and caused the production of flour in the Minneapolis plant to increase from 2 million barrels in 1880 to 16 million in 1910: "It was the roller mill that wrested the milling crown from Budapest in Europe and St. Louis in [the] United States and gave it to Minneapolis...and was responsible for an agricultural as well as an industrial evolution" (Peterson 1976). The agricultural revolution took place because roller mills were able to grind the hard and nutritious spring wheat more efficiently than querns.

CONCLUSION

The history of size reduction in the grain-milling industry parallels the growth in the development of processes, machinery, and technology, as more sources of energy became available to replace manual energy in driving the machinery. Although size-reduction techniques differ among the various industries, which use the process best suited to the particular needs of each, the pattern of growth in all the industries is the same. By the end of the 19th century, all the elements of modern-day grain milling were in place, and, by 1930, the gradual reduction system was in universal use. In the years that followed, the rolls were continually improved, with, for example, high-pressure hydraulic systems for roll support, better bearings, and special wear-resistant roll surfaces and liners.

CHAPTER 5

Stamp Mills and Crushers

Until the 15th century, all the major developments in size reduction were for the food preparation industry–for the grinding of grain to make flour. Although there were demands for minerals, particularly gold, copper, zinc, and lead, and for construction materials, the demands were small, and the methods used to grind grain were adequate. When manually operated tools and machines, however, were applied to crushing and grinding rocks and mineral ores, the size-reduction process took longer because the materials were much harder and required more energy to break. Rocks and mineral ores were also much larger than grains and had to be broken manually before they could even be fed into a mortar and pestle or a quern. Finally, rocks and mineral ores are much more abrasive than grains, increasing the wear and tear on and decreasing the life span of the crushing and grinding equipment.

The increased demand for metals, cement, and aggregates, which began around the 15th century, reached its peak during the 19th and 20th centuries. To meet the increased demand for machinery that could crush and grind metallic and industrial mineral ore, cement raw material and clinker, and large rocks to the smaller size needed for aggregates, new classes of heavy-duty equipment using emerging sources of energy were required. As discussed in the last chapter, first waterpower was used with water wheels to drive the machinery. Next came steam engines, which led to further size-reduction developments. Then, electric motors replaced steam engines, reenergizing the progress of size-reduction technology once again. It is interesting to note that water, and the steam made from water, is the main source of energy that drives the turbines that drives the electric generators.

Because rocks, cement raw materials, cement clinker, and mineral ores all require the same basic types of crushing and grinding equipment, we devote this chapter and the two that follow to stamp mills and crushers, roller mills, and tumbling mills. We discuss these technologies roughly in chronological order of their development and also present information on the uses for each.

STAMP MILLS FOR ORE BREAKAGE

As the world's population and its demand for minerals continued to grow, it became necessary to devise a machine that could crush abrasive ores to small sizes at higher rates and that could be driven using a renewable power source. Enter the stamp mill.

Although water-driven trip hammers had been used in China for crushing ore pebbles since 300 AD (Figure 5.1), their capacities were very limited. Beginning about 900 AD, dry stamp mills driven by muscle power (Figure 5.2) had been used to make iron, and, by the 12th century, they were being used to a minor degree for liberating minerals from ore for gravity concentration. The early dry-stamp mills were essentially mechanized mortars and pestles.

The woodcuts in *De Re Metallica* (Agricola 1950), reproduced in Figure 5.3, show the stamp mill and water wheel with an operator wheeling a barrow of coarse pebbles

FIGURE 5.1 Water-driven trip hammer. Filling the bowl with water caused it to fall and raised the hammer. When the bowl tilted enough, the water discharged and the hammer fell (Reprinted by permission from Cambridge University Press)

FIGURE 5.2 Dry stamp mill operated by muscle power (Cook 1974; reprinted by permission from Dyno Industries)

A

B

FIGURE 5.3 (a) Stamp mill (Lebeter 1949a) (b) quern shown in woodcuts in *De Re Metallica* (Agricola 1950)

back to the mill for further grinding, and the quern driven by the water wheel that grinds the stamp mill product. The woodcuts evoke an era of grinding in which the working conditions were relatively pleasant compared with the grim conditions of earlier years.

The early dry-stamp mills were used to grind and mix sodium or potassium nitrate, sulfur, and carbon to make black powder, which was a widely used explosive.

By 1500, the Renaissance was in full swing, and with it came a renewal of the culture of arts and sciences that had flourished in Greek and Roman times. The intellectual vitality

A

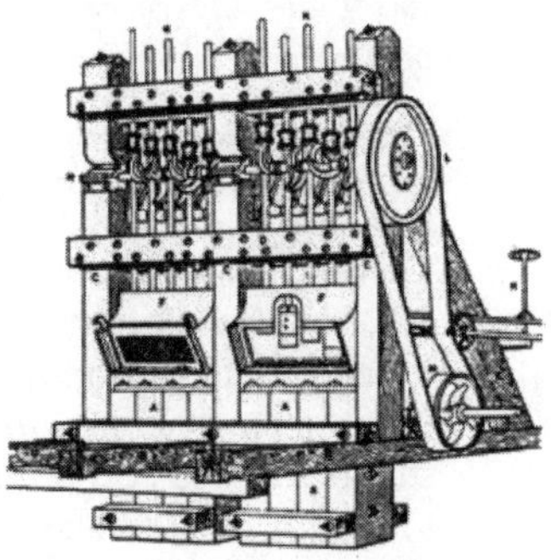

B

FIGURE 5.4 (a) Water wheel used to drive a stamp battery in Castlemaine, Victoria, in the 1880s (reprinted by permission from Castlemaine (Victoria) Art Gallery and Historical Museum) (b) 1880 stamp mill (Lebeter 1949a; reprinted by permission from *Mining Journal* [formerly *Mine & Quarry Engineering*])

of the Renaissance created an environment that encouraged new ideas and radical improvements to current ideas. The resulting prosperity created a high demand for metals, and the production of silver and copper in central Europe increased fivefold between 1460 and 1530 (Nef 1958). This required a higher capacity for grinding ores than could be provided by manually operated machines, and the sources of renewable energy, such as water and wind, that were being used for flour milling were applied to mineral liberation as well.

Sigismund von Maltitz introduced wet stamp mills in 1512 (Scheibe 1993), and eventually it was common for ores to be ground wet in stamp mills where slurries containing fine particles would overflow readily from the mills, allowing them to be operated continuously. It was the water-driven stamp mills and the iron-shod stamps that made possible the development of high-capacity ore-processing systems, such as those in the mining districts near Schwaz in Tyrol in the Alps and in the main centers for silver and copper mining in Europe.

Stamp mills powered by water wheels dominated fine crushing and grinding from about 1500 until the first half of the 20th century. Figure 5.4 shows a water wheel of the type used to drive a stamp mill in the 1880s, along with a large stamp mill of the time. According to Stone and MacKinnon (1976), "The wheel was 72 feet in diameter with 220 buckets around the circumference. Each bucket held 5 gallons of water and the wheel supplied power to the stamp battery through a 38 foot gear wheel."

Stamps were the only machines available that could continuously grind ores to finer than 1 mm, a suitable size for gravity concentration. The gravity concentration process liberated gold, lead, and copper minerals from the relatively coarse-grained ores mined at the time. The very fine particles produced in the stamp mills, however, could actually interfere with gravity concentration. In addition, stamp mills were large, heavy, difficult to maintain, and noisy, creating vibrations throughout the area where they were installed. The mills were not easy to operate efficiently, and T.A. Rickard (1897) wrote with his usual wit about how some manufacturers sought to take advantage of this:

> *There are establishments which have what they call a standard type of mill which they highly recommend for the reduction of ores running through a whole gamut of differing compositions. Like the iron bed of Procrustes, to which the wayfarer had to suit his length at the risk of summary abbreviation or painful elongation,*

> *so the manufacturer expects the ore to adapt itself to his mill or choose between being labelled unprofitable or refractory.*

In 1871 Mark Twain worked as a stamp mill operator and wrote a memorable description of the machine (Paul 1970):

> *This mill was a 6-stamp affair, driven by steam. Six tall, upright rods of iron, as large as a man's ankle, and heavily shod with a mass of iron and steel at their lower ends, were framed together like a gate, and these rose and fell, one after the other, in a ponderous dance in an iron box called a battery. Each of these rods or stamps weighed 600 pounds. One of us stood by the battery all day long, breaking up masses of silver-bearing ore with a sledge and shovelling it into the battery. The ceaseless dance of the stamps pulverized the rock to powder, and the stream of water that trickled into the battery turned it to a creamy paste. The minutest particles were driven through a fine wire screen which fitted close around the battery and were washed into great tubs warmed by superheated steam–amalgamating pans they are called.*

Stamp mills reached their peak efficiency, with production rates of up to 200 tpd, late in the 19th century in the United States and Australia. Large ore-processing plants contained more than 150 stamps. Ores containing coarse gold, which were mined during the boom years in California and Victoria, Australia, starting in 1850, were soon depleted, and operators needed to process low-grade, fine-grained gold ores. Gold ores in Colorado and California were examples of ores of similar hardness but different compositions that required different grinding in stamp mills. Table 5.1 shows the performance of stamp mills on these ores in 1885. Optimizing the operation of stamp batteries and the associated amalgamation processes was no easy task, and skilled engineers were required to operate these circuits.

In the late 19th century, stamp mills were used prominently in the gold fields of South Africa, before the advent of the pebble mills. Stamp mills, however, were limited by their production capacities and the product size that was economically possible. But for all their problems, stamp mills were the main fine-crushing and coarse-grinding machines available from 1500 to 1900, and they rank as one of the great inventions for size reduction. Stamp mills lost favor when fine crushers and ball mills–which were less expensive and easier to maintain and operate–proved to be reliable as continuous operating units. By 1915, new installations of stamp mills were scarce, ending an era that had lasted 400 years.

THE ADVENT OF STEAM POWER

Every revolution requires a catalyst, and it was the need to pump water from mines that sparked the revolution of steam power. By 1650, coal and tin mining were important industries in England, and water had become a serious problem in many mines that had been worked for centuries, particularly the deep tin mines in Cornwall. Pumps required more power than could be supplied manually if the mines were to be kept in operation.

The idea of using steam as an energy source was not new; the Italian physician Branca had used steam for grinding in 1629, making the steam in a metal boiler shaped like a man's head. The only place the steam could get out was through a tube held in the man's mouth, from which a jet of steam issued and blew against a wheel, making it turn. The wheel was geared to a shaft lifting two rods that Branca used for crushing medicine in a mortar (Hartman 1940; see Figure 5.5).

The principle of steam power is that when 1 L of water is boiled, it becomes 1,849 L of steam. If a cylinder containing a piston is attached to a boiler, the conversion of water

TABLE 5.1 Performance of stamp mills on ores in Colorado and California

Ore Type and Performance Measure	Colorado	California
Ore type	Fine grained—15% pyrite	Coarse grained—2% pyrite
Weight per stamp, kg	250	386
Height of drop, cm	46	13
Drops per minute	30	90
Bottom of screen to top of die, cm	36	10
Screen aperture, mesh	50	30
Product size for a 40-mesh screen	70%, 100 mesh	50%, 100 mesh
Energy used per stamp, kW	0.75	0.75
Tons per hour per stamp	1	3

Source: Rickard 1897.

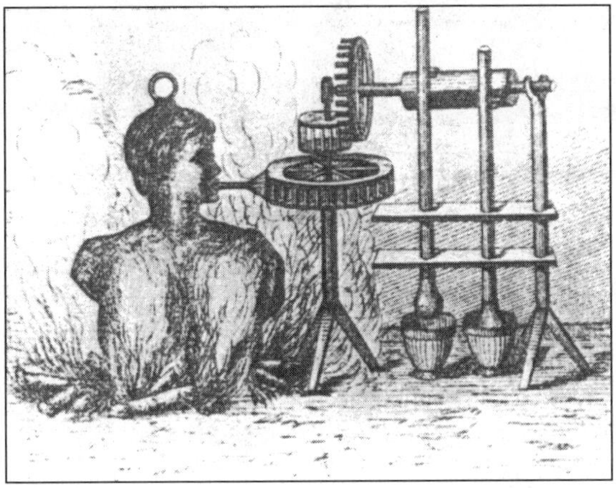

FIGURE 5.5 The Italian physician Branca invented a steam engine that crushed medicine in a mortar in 1629 (Hartman 1940)

in the boiler to steam will drive the piston to its limit; it will return to its original position when the steam is condensed and a vacuum is formed. Eventually a machine known as a beam engine, shown in Figure 5.6, was built using this principle.

The beam engine consisted of a beam pivoted in the center with one end attached to a piston that moved up and down as fresh steam was introduced and then condensed, and the other end attached to a chain and bucket that could be used to lift and discharge water every cycle. Thomas Newcomen built the first commercially successful beam engines, and by 1729 more than 100 were in operation at coal mines. The loss of heat by the piston and cylinder during every cycle, however, resulted in high fuel costs. In 1765, James Watt modified the Newcomen engine by including a separate condensing chamber. Eventually, Watt formed a partnership with Matthew Boulton, who built steam engines to Watt's design that reduced the cost of fuel for pumping by two-thirds. Beam engines using steam engines of Watt's design were soon installed at many mines. Operators realized that this source of power could be used for other purposes, including size-reduction machinery, and steam engines came to be used to drive stamps and crushing rolls.

Although steam-powered drills were available by 1860, it was difficult to use steam power underground. The Walhalla mine in Victoria, Australia, was a rich gold mine from which 14 tons of gold were extracted starting late in the 19th century. Because the siliceous ore was very hard, the owners chose to use steam-powered drills. They installed

FIGURE 5.6 Appearance and method of operation of the steam-driven beam engine (de Camp 1961)

six large boilers in a huge cavern underground, which required several tons of wood each day to keep the boilers working. It was only the size and wealth of the deposit that made steam-powered drilling possible.

Steam power gave a new direction to inventors interested in crushers and changed ideas about mechanical crushing. Unlike water and wind, steam was an intense and reliable source of energy that could be used anywhere at a controllable rate.

Crushing is the low-energy phase of the comminution process, and grinding is the high-energy phase. Crushed products for direct use are known as aggregates, which are hard inert materials used for mixing various-sized fragments with a cementing material to form concrete or mortar. Aggregates are used in commercial centers, homes, airports, highways, dams, and streets. When aggregates from natural deposits were not available, they were made from rocks by hammer and chisel, or sometimes even in a mortar and pestle, and later in crushers developed for the purpose.

Before 1855, when the jaw crusher was invented, the hammers, drills, and chisels used to break rocks were powered by hand. The broken rocks were sorted into the sizes needed for various grades of aggregates on stationary screens set up at an angle and moved in wheelbarrows and handcarts (Figure 5.7). Screening surfaces were stretched over wooden or metal frames, and the material to be separated was shoveled onto the screening surface by hand. The original surfaces were made of heavy-duty cloth, but the advent of metal wire woven to a specific opening size and then of steel plates punched with sized openings extended the lifetimes of screening surfaces. In addition, operators discovered that, if the screens were vibrated, the flow through and over the screen improved, making vibrating screens a vital part of aggregate production.

In the 19th century, specifications were developed for the size distribution and shape of aggregates. When the automobile powered by an internal combustion engine was invented, more and better roads were needed. Beginning in about 1815, a construction system developed by John MacAdam was used for building roads. In this system, which became known as "macadam" through usage, the surface of a road was formed by interlocking stones that had passed through 50-mm apertures and then sealing the surface.

FIGURE 5.7 Manual handling of ore after blasting in a quarry

The specifications for aggregates became explicit, particularly when portland cement–made by burning a mixture of lime and clay–began to be used to make concrete for main roads. Soon asphalt became the main surfacing material (although, where the climate permits, cities now use concrete as the top surface for streets), and, with the invention of the airplane and the subsequent increases in the number, weight, and size of airplanes, sturdy airport runways became necessary. The first surfaces of airport runways were gravel (small aggregate), and later gravel was surfaced with asphalt. Today, runways are gravel covered by reinforced concrete. The requirement for efficient crushing was a hallmark of all these processes.

Cornish Rolls

In Cornwall in 1804, John Taylor at Wheal Travistock set up two discarded pump barrels in parallel (Barton 1961) as horizontal rolls to crush copper and tin ores for concentration by gravity. This experimental crusher, which became known as "Cornish rolls," was driven by a steam engine (Figure 5.8). The rolls were only one of the many new methods that Taylor, a distinguished scientist who had been elected as a Fellow of the Royal Society, introduced. His advanced ideas of management and technical and financial control of metal mines transformed the industry (Lynch 2002).

The Cornish roll crusher was very successful and soon saw widespread use, making an important contribution to the size reduction of materials. But there was still much to be learned about crushing:

> *The whole secret of the efficient reduction of any material lies in a suitable crushing unit of the right capacity. This, I suppose, is obvious, but when one looks at some of the reduction plants in operation one is inclined to the belief that the obviousness of such a statement has escaped at least a few. The best way to reduce any material is also the cheapest: the energy must be proportioned to the work in hand, otherwise either the work is not being done properly or energy is being wasted. (Seymour 1924)*

During the second half of the 19th century, inventors responded with jaw and gyratory crushers that are still used today, among them the Blake jaw crusher that was patented in 1858 and the Gates gyratory crusher that was patented in 1881.

By 1850, smooth double rolls for grinding ores driven by 15-kW motors were operating at 50 rpm and grinding 60 tpd of siliceous ores. These rolls had come a long way from the ones that Taylor used in Cornwall in 1805. From 1890 to 1910, the cyanide and flotation processes were established (see Chapter 7), and reliable fine crushers were

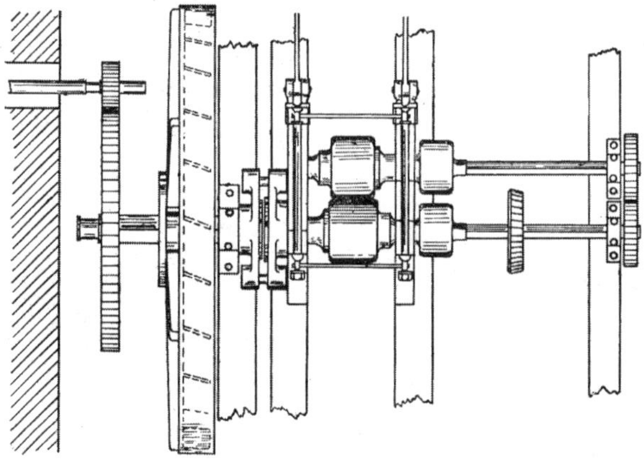

FIGURE 5.8 Cornish rolls (Lebeter 1949b)

required to grind ores at high rates to less than 5 mm, which was suitable then for ball mill feed. Roll crushers were the answer, and models were built that were stronger and more compact than those previously available and operated at higher speeds. By 1925, rolls were 1.8 m in diameter and 0.53 m wide and were powered by 112-kW motors. They typically ground siliceous ore to −3 mm.

Figure 5.9 shows the type of roll crusher that was in common use worldwide for most of the 20th century. Roll crushers used two steel rolls that rotated toward each other, exerting high pressure on the material passing between them. One roll revolved in fixed bearings, the other in a movable bearing-and-bracket assembly held in position by high-compression springs. The ore was fed into the wedge formed by the rotating rolls and was pulled into the breakage zone by gravity and friction between the ore and the face of the rolls until it passed through the shortest gap between the rolls. In the early part of the 20th century, the pressure on double-roll crushers was limited to 70 tons per roll by the materials of construction that could be used for the springs and by the design procedures available.

With the development of hydraulic pistons, much higher pressure became available, but the demand for double-roll crushers had decreased by then, because rod mills and very fine cone crushers had higher capacities and less downtime than the double-roll crushers.

To obtain a consistent product size, it was important to maintain an even wear across the face of the rolls. This required maintaining a uniform particle-size distribution and thickness of the stream of feed across the face of the rolls. Without these, wear patterns across the face would be uneven, leaving an uneven gap between the roll faces and resulting in an adverse coarsening of the crusher product. Also, tramp steel could gouge the face of the rolls, which would also affect the size of the product. Generally, the maintenance on roll crushers was high, both on the roll faces and on the side wear plates.

Rolls were successfully used for fine crushing in aggregate plants until the 1920s, when they began to be replaced with fine-crushing modified gyratory crushers and then by cone crushers.

Blake Jaw Crushers

The Blake crusher had its origin in a decision by the town council in New Haven, Connecticut, in 1852 to build a 3.2-km road in the center of the town using the MacAdam process. Eli Blake was one of the townsmen responsible for planning the road, and he realized that the cost would be reduced if a machine could break the quarried rock faster

FIGURE 5.9 Typical roll crusher (Richards and Locke 1940)

and cheaper than the slow and costly hand-breakage method. The crusher he invented was accepted rapidly by the stone industry and was said "...to do the work of 100 men in a given time and at less than one third the cost" (Dickinson 1945). Blake's inventive brilliance is shown by the fact that Blake-type jaw crushers are still widely used today and the design has changed little in its essentials from his original design.

> *Eli Blake ...conceived the stone breaker in all its parts, even to their weight and strength, before ever a machine was made. (Dickinson 1945)*

> *All of the large, heavy duty primary crushers of the jaw type are built around the Blake principle which, for simplicity and brute strength, is unsurpassed by any mechanism thus far devised for rock and ore breaking. (McGrew 1950)*

The jaw crusher Blake designed in 1853 and patented in 1858 was one of the great inventions of size-reduction machines (Dickinson 1945). The crusher, shown in Figure 5.10, had a fixed jaw and a moving jaw that converged at the discharge end. The moving jaw pivoted on a shaft at the top, its base had a reciprocating movement of about 12.5 mm, and it was driven through a toggle. The toggle joint applied immense force to a rock held between the jaws as it straightened and shattered any brittle rock, irrespective of its hardness. A fly weight could be added to provide extra force to push the moving jaw through the crushing zone.

The early Blake crushers operated at 5–15 tph depending on the size of product, but their capacity increased considerably as larger units were built. Although they were used extensively for road building, it was 1861 before they were used to crush ore. But the immediate success of the Blake crusher in a plant on the Comstock lode in Nevada dispelled all hesitation about using them in mineral processing plants. The Blake jaw crusher was the first success that emerged from the plethora of inventions in the 19th century for size reduction of rocks and pebbles.

Gates Gyratory Crushers

Philetus Gates was one of several inventors who patented rudimentary designs incorporating the idea of a circular crusher between 1860 and 1878, but it was not until 1881 that he patented a successful gyratory crusher (Dickinson 1945). Figure 5.11 shows how the original Gates crusher worked, and early models are shown in Figure 5.12.

The Gates crusher consisted of a vertical cone suspended at the top and held by an eccentric sleeve at the bottom that gyrated within a fixed cone. Crushing occurred around the periphery as the movable crushing surface, named the mantel, approached to

A B

FIGURE 5.10 Blake jaw crusher: (a) general view (b) mechanism (Lebeter 1949b; reprinted by permission from *Mining Journal* [formerly *Mine & Quarry Engineering*])

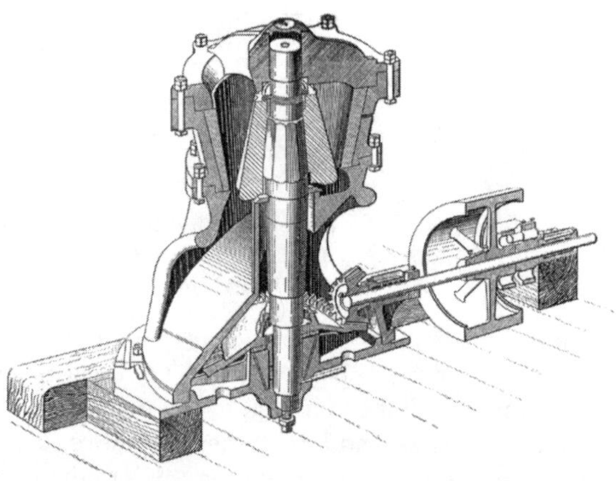

FIGURE 5.11 Gates gyratory crusher (Louis 1894)

and receded from the fixed crushing surface, named the concave. The circular top gave a larger feed opening and reduced blockage. The first sale of a Gates crusher predated the patent by several months when a No. 2 crusher was sold to the Buffalo Cement Company in New York in 1880. That was the first of several thousand gyratory crushers that carried the name of Gates to the far corners of the earth (McGrew 1950).

The Blake and Gates crushers were rival machines, and the custom of the day was to conduct trials between competing pieces of machinery and award the sales contract to the winner. In 1883 in Meriden, Connecticut, a contest was staged between a Blake jaw crusher and a Gates gyratory crusher. Each machine was required to crush 7 m^3 of stone, with feed size and discharge settings being similar. The question of productivity was quickly settled when the Gates crusher finished its quota in 20.5 minutes and the Blake crusher in 64.5 minutes. "This must have been a sad disappointment to the proponent of the Blake machine who happened to be the challenger" (McGrew 1950).

The Blake jaw crusher and the Gates gyratory crusher are still the preferred primary crushers today, and their designs have changed little from the originals. The Gates crusher soon became very popular. During the late 19th and early 20th centuries, the

A B

FIGURE 5.12 Gates crushers: (a) built in 1880 and (b) in 1910 (McGrew 1950; reprinted by permission from Primedia)

trend was to increase crushing capacity by duplicating small units. "In 1915 at the huge Homestake gold ore mine in South Dakota there were no less than 22 small Gates gyratory crushers to prepare the ore for the batteries of stamps containing a total of about 2500 stamps" (McGrew 1950).

The development of jaw and gyratory crushers had a major impact on large mining operations. In many large open-pit mines, steam shovels were used to fill the ore cars with ore that had been blasted from the face with dynamite, and steam locomotives pulled the loaded cars to the primary crusher. Similar loading and transport systems were used for major civil engineering works, and photos of the Panama Canal show the steam shovels and rock trains used in its construction. The early large open-pit mines such as Utah Copper's Bingham Canyon mine near Salt Lake City, Utah, and the Hull Russ Iron Ore pit near Hibbing, Minnesota, incorporated steam shovels and railroad transport systems in their original operations. For their driving energy source, the shovels went from steam engines to internal combustion motors and then to electric motors.

Edison Giant Rolls

In 1895, Thomas Edison was responsible for a remarkable innovation in crushing. Edison believed that high-grade deposits of iron ores in the eastern United States were running out, and he decided to develop a deposit of low-grade ore containing magnetite at a treatment rate of up to 6,000 tpd. Minimizing mining and crushing costs was essential to success, and the high capital and operating costs of multiple small crushers was a particular area of concern. He reasoned that the recoverable energy in 1 lb of coal and 1 lb of 50% dynamite was approximately the same but the cost of the dynamite was 100 times the cost of the coal (Dyer and Martin 1910). Consequently, it would be much cheaper to break large pieces of ore by machines driven by steam power than by explosive energy, particularly because much of the dynamite was used for rebreaking rocks from the first explosion. His approach was to break the rock by dynamite into large boulders—up to 5 to 6 tons—and then use huge crushers to break these into small particles.

> *Edison conceived the bold idea of constructing gigantic rolls which, by the force of momentum, would be capable of crushing individual rocks of vastly greater size than ever before attempted. ...The giant rolls were two solid cylinders, six feet in*

> *diameter and five feet long, made of cast iron. To the faces of these rolls were bolted a series of heavy, chilled iron plates containing a number of projecting knobs two inches high. Each roll had also two rows, of four inch knobs, intended to strike a series of hammer-like blows. The rolls were set face to face fourteen inches apart in a heavy frame, and the total weight was one hundred and thirty tons of which seventy tons were in moving parts. The space between these two rolls allowed pieces of rock measuring less than fourteen inches to descend to other smaller rolls placed below. ...Previous to the dumping of a skip the rolls were speeded up to a circumferential velocity of nearly a mile a minute, thus imparting to them the terrific momentum that would break up easily in a few seconds boulders weighing five or six tons each. This sudden strain might have tended to stop the engine driving the rolls but by an ingenious clutch arrangement the belt was released at the moment of resistance in the rolls by reason of the rocks falling between them. (Dyer and Martin 1910)*

Eventually 2.4-m × 2.1-m Edison rolls were built, and they must have been spectacular to watch. But their costs of operation would have been high, and they soon declined in popularity. The giant rolls were only one of many innovations in the complex mining and processing system devised by Edison that resulted in very fine magnetite particles being extracted from low-grade ores. The rolls were a heroic effort by Edison to reduce crushing and grinding costs to a manageable figure, but ultimately the process was not an economic success. Edison rolls created much interest in the mining industry for a while, and several were installed in other plants, including four sets operating in series on quarried limestone in Edison cement plants. But jaw crushers and gyratory crushers soon proved to be superior, and the giant rolls passed into history.

The crushers that were available in the late 19th century could be fully used in mines only if the ore loading and transport systems from the mine or quarry to the processing plant could keep up with the capacity of the crushers. By then open-pit mining and quarrying had become the predominant mining method, but handwork was a serious limitation to efficiency. At the end of the 19th century, steam power was widely used for loading and transporting ore and rocks, although manual handling continued far into the 20th century.

Beginning in the 1920s and 1930s, large internal combustion engines were developed that could drive large trucks. After World War II, ore trucks began to replace railroad transportation of ore from open-pit mines. In the latter part of the 20th century, primary crushers were built on tracks located in open-pit mines and quarries and moved as the face of the mine moved. Ore could either be dumped by the shovel directly into the primary crusher or could be hauled by truck from the shovel to the primary crusher. The primary crusher discharge was conveyed out of the pit and to the processing plant on wide-belt conveyors. Overburden was also conveyed from the mine face to the overburden dumping area.

CRUSHERS IN THE 20TH CENTURY

The incentive to build large primary gyratory crushers came in the 1920s with the development of open-pit mines for copper and iron ores. The aggregate producers' demand for a machine that would control the size distribution and shape of pebbles at a high production rate offered the incentive to build cone crushers, which had specially designed crushing chambers to minimize blockage. In a cone crusher, the closed side settings could be adjusted to control the product size during operation. When necessary, these machines could be made larger and stronger to crush hard, abrasive ores of metallic minerals for further grinding. Limestone used for making cement is usually crushed with a jaw or gyratory crusher in the quarry and a cone or impact crusher in the plant ahead of the roller or ball mills.

Primary Crushers

By 1910, the need to crush high tonnages of rock from open-pit mines led to a spectacular growth in crusher size. In 1905, a large jaw crusher was 0.77 m × 0.61 m and was driven by a 34-kW motor, and a large gyratory crusher was 0.46 m × 1.61 m with a 75-kW motor (Louis 1909). Within 15 years, a large jaw crusher was 2.15 m × 3.07 m and a large gyratory crusher was 1.84-m. Both machines required 375-kW motors (Taggart 1927). The growth of copper mining led Allis-Chalmers to build six all-steel gyratory crushers with 1.54-m feed openings for a copper mine in Chile in 1926–1927. They weighed 500 tons each and were the height of a two-story building. Figure 5.13 shows one of these crushers.

The increasing demand for iron also affected crusher design. The high-grade pockets of hematite in the Iron Range in Minnesota were depleted during the world wars, and the underlying low-grade taconite ore, which is a hard, abrasive siliceous ore containing magnetite, had to be mined. Larger and stronger crushers were needed, and Traylor, Nordberg, and Allis-Chalmers all built high-capacity, heavy-duty primary and secondary crushers that incorporated improved lubrication systems and more reliable drives and bearing elements. The maximum power that could be drawn by the improved 1.54-m primary gyratory crushers was increased to 750 kW.

By 1970, crusher plants were operating that handled more than 100,000 tpd, and by 2000, plants could be built to crush any tonnage of ore. These plants are immense structures that require deep excavations to achieve good gravity flow of the pebbles. Figure 5.14 is a sketch of the primary crusher at Bougainville Copper Limited (Papua New Guinea), which handled 120,000 tpd of low-grade copper ore. Developments on Bougainville Island added much to knowledge of crushing systems.

Single-Roll Crushers

Several types of single-roll mills have been built in which the solids are broken by being nipped between the rotating roll and a fixed surface. An early single-roll machine was the sledging roll crusher, which was built by Allis-Chalmers in 1911. It was a single-roll toothed crusher that was typical of a family of similar crushers built during the 20th century. Two units (0.9 m × 1.5 m) were installed in the Fairmount, Illinois, flux limestone plant of the Casparis Stone Company, and they quickly achieved a high degree of popularity. Units of this type came to be called "Fairmount" crushers, and the line was expanded to include smaller sizes as well as a big 1.5-m × 2.1-m machine, which is shown in Figure 5.15.

> *The roll center is made of cast steel and the teeth are of manganese steel inserted in cored pockets in the surface of the roll center. These teeth, it will be noted, are of different heights, the higher teeth being known as the slugger teeth and the lower ones as the regular teeth. The tip velocity of the slugger teeth in the Fairmount crusher is from 400-450 ft per minute. The working faces of the teeth are radial in profile which means that the faces are normal to the line of action when a rock is contacted in any part of the crushing chamber. ...On the soft and medium grades of stone, to which the single roll crusher is particularly well adapted, the machine will turn out a product carrying less fines than will either the jaw or gyratory types. ...most of the crushing is done by clean sharp sledging blows against the individual pieces. (McGrew 1950)*

A

B

FIGURE 5.13 (a) All-steel Allis-Chalmers gyratory crusher with a 1.54-m feed opening built for a copper mine in Chile (Peterson 1976; reprinted by permission from Milwaukee County Historical Society) (b) feed opening of the crusher. The size of the machine can be judged from the sizes of the men (McGrew 1950; reprinted by permission from Primedia).

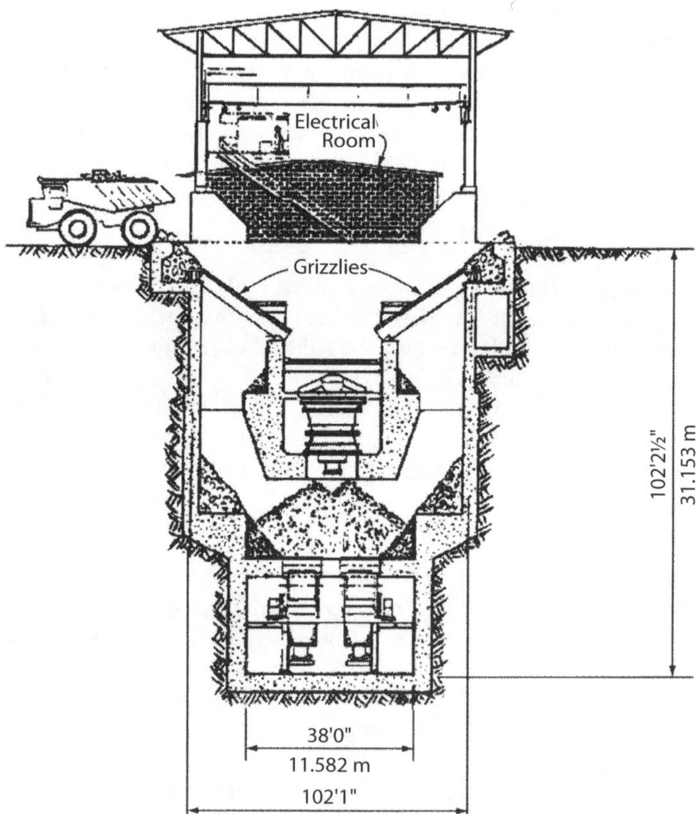

FIGURE 5.14 Sketch of the primary crusher installation at Bougainville Copper Limited (McQuiston and Shoemaker 1978)

FIGURE 5.15 Fairmount single-roll crusher (McGrew 1950; reprinted by permission from Primedia)

Similar types of single-roll crushers were made by many companies during the 20th century. Because of their low fines production, these crushers were suitable for coal breakage. Coal, which is brittle and breaks easily, had been broken manually until about 100 years ago, when engineers began to adapt hammer mills and roll crushers for crushing coal, and low- and high-speed crushers were used to crush run-of-mine coal to about 150 mm.

Secondary Crushers

Secondary crushers were not used until the large open-side settings in the huge primary crushers built early in the 20th century made them necessary. The first secondary crushers were scaled-down standard gyratory crushers, but they had a tendency to “choke” and their productivity was low, making them inefficient for fine crushing. The growing popularity of reinforced concrete created an increasing demand for smaller sizes of crushed stone and gravel, and attempts to meet this demand with existing crushers highlighted the need for a better fine crusher. The crushing system in a stone plant built in Thornton, Illinois, in 1913 was typical of the new development. It comprised one 1.2-m primary crusher, four secondary crushers with 178-mm feed openings, and 16 finishing crushers—all gyratories. Two sets of roll crushers were added later to augment the production of small stone.

Attempts were made before World War I to adapt the standard gyratory to fine-crushing duties by making a “shorthead” model, which consisted simply of an abbreviated crushing head installed in a standard machine with concaves to match. Crusher liners were known as “concaves,” because they were attached to the inner surface of the crusher forming the concave crushing zone. This model did not prove to be very successful; crushing stresses were concentrated at a point where the top shell was ill-equipped to withstand them. In addition, the throw at the point of discharge was too small to take full advantage of the increased diameter of the discharge opening.

By the early 1920s, several engineering companies were changing the design of the crushing chamber and the concaves in order to build efficient fine crushers. The Superior–McCully fine-reduction crusher was the first to show a significant improvement over other models when the crushing chamber was modified to include a cylindrically bored shell, vertical concaves, and a flared crushing head. The eccentric speeds were increased, and throws were adjusted for effective operation at fine settings. Then the concaves were tapered at both ends to distribute the wear better, and eventually these “nonchoking” concaves became standard. This development of nonchoking concaves, using the principle of curved-profile crushing chambers, was the most important and far-reaching improvement in crusher design that had been made for many years—possibly the greatest since the inception of the gyratory crusher. Allis-Chalmers designed nonchoking concaves that improved the productivity of crushers and had the advantage that they could be fitted to hundreds of crushers existing at the time. Figure 5.16 shows the differences between straight-faced concaves and nonchoking concaves.

The curved nonchoking concaves crushed rocks more effectively before they entered the final crushing region, and they controlled the flow of pebbles through the crusher better. Allis-Chalmers' strategy was to maintain its position as a leading manufacturer of size-reduction equipment by innovative engineering and design work, and by the occasional purchase of competitors such as the manufacturer of the Superior–McCully fine crusher and its line of primary gyratory crushers known as the McCully crusher. Soon after this purchase, Allis-Chalmers was offered the patents for a reduction crusher developed by Will Symons, a Chicago engineer. Allis-Chalmers declined, however, no doubt because two new and competing crushers within a short time would have been difficult for the company to digest. Instead, the Nordberg Company of Milwaukee purchased the patents for the Symons high-speed secondary crusher. In time the Symons cone crusher virtually killed the sales of the McCully fine-reduction crusher and seriously reduced the sales of the smaller Allis-Chalmers crushers.

Cone Crushers

Compared to the gyratory crusher, the cone crusher is characterized by its higher speed and a flat crushing chamber design, which is intended to give a high capacity and reduction ratio for materials suitable for this type of processing. The aim is to retain material longer in the crushing chamber to do more work on material as it is being processed (SME–AIME 1985). Figure 5.17 shows how stone flows through the crushing chamber of a cone crusher.

All crushers developed up to the 1920s, except for the very large gyratory crushers, were provided with some means of adjustment to compensate for wear or to adjust for variations in particle size. Generally, adjustments had to be made when the crusher was off line, but this was not a problem with machines that were used for coarse crushing to about 50–100 mm because the exact setting of the discharge opening was not critical and wear was slow. However, when it became necessary to crush to 10–25 mm, the setting of the smaller discharge opening had to be maintained within close limits, and the crusher was designed so that it could be adjusted mechanically without having to stop to reset the wearing parts. The design of the crushing chamber was also modified to suit the task. The fine-crushing Symons shorthead crusher has a chamber that is shorter and of slightly different shape than the standard coarser crusher, as shown in Figure 5.18.

The cone crusher has been widely used in the mining industry for 70 years to prepare feed for rod and ball mills and to crush critical-size pebbles that have been removed from autogenous or semiautogenous (SAG) mills because they are too small to be effective as grinding media and too large to be broken by the larger rocks. Its main use has been in the crushed-stone industry for producing aggregates to meet the demands set by the construction industries. Automated control systems have been developed to ensure that specifications for the various products are met.

Impact Crushers

In these types of crushers, high-velocity impacts are used to break rocks. They work either by hitting rocks with a high-speed hammer (hammer mills) or throwing rocks against each other or against a steel plate (Barmac crusher). Breakage by impact was the

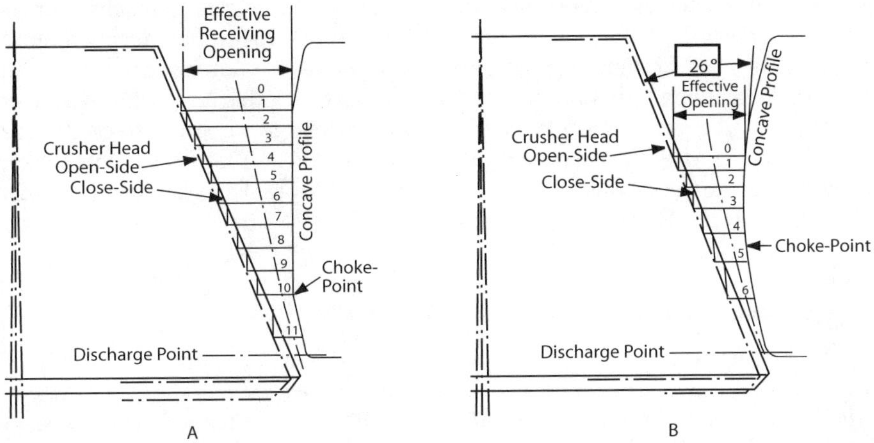

FIGURE 5.16 Crushers with (a) straight-faced and (b) nonchoking concaves (McGrew 1950; reprinted by permission from Primedia)

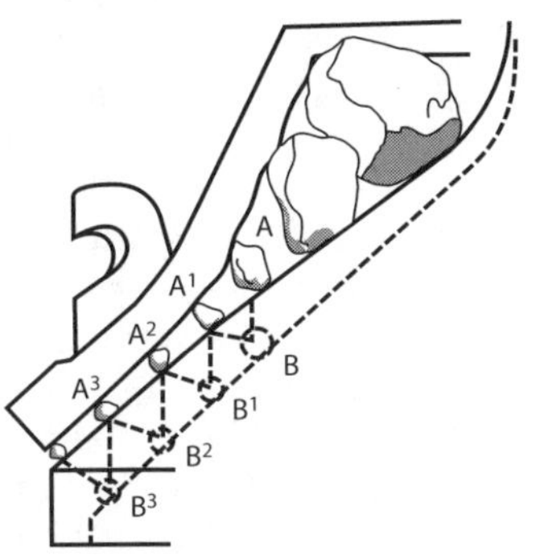

FIGURE 5.17 Material flows through the crushing chamber of a cone crusher (SME–AIME 1985)

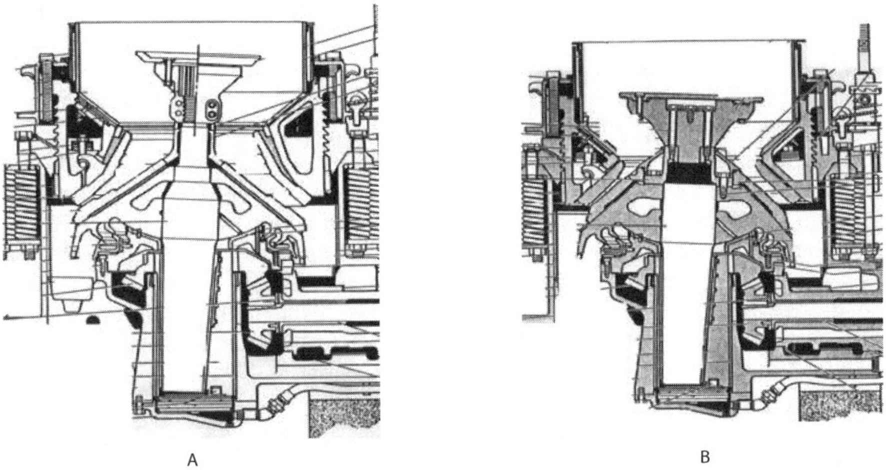

FIGURE 5.18 Crushing chambers of the (a) Symons (Nordberg) standard and (b) shorthead cone crushers. The different shapes of the chambers can be seen (courtesy of Metso Minerals).

concept behind the Hadsel autogenous mill, discussed later, but in most cases breakage by impact did not work well in the Hadsel mill because the rock velocities were not high enough for them to be broken to small particles. It was eventually realized that shattering coarse rocks by impact in an autogenous mill was the opposite of what was needed. To prevent the mill from being filled with pebbles too large to be discharged from the mill, slow breakage of the coarse rock by abrasion with the finer ore being broken by impact by the coarse rocks was necessary. When impact breakage is required, it can be achieved and controlled in impact crushers.

Hammer Mills

In 1830 a patent was issued in United States for a machine

> *...which comprised a wooden box containing a cylindrical drum, apparently made of wood, on which iron knobs or hammers were fastened. The expectation was that this drum, when revolved at 350 rpm, would shatter the rock fed into the box. (McGrew 1950)*

This machine never went into commercial production, but it was the forerunner of the hammer mill, which is the most widely used crusher in which high-velocity impacts are used to break rocks. The distinctive features of the hammer mill are the high shaft speed, which may be in the 700–1,500-rpm range. The high velocities of the tips of the hammers–up to 3,000–4,000 m/min–gives them sufficient kinetic energy to shatter rocks at the instant of impact. These velocities could be achieved with electrical motors more easily than with steam engines, so hammer mills had to wait for electricity in the 20th century to be widely used for breaking hard rocks even though they were used in the 19th century. Figure 5.19 gives a sectional view of the operation of a hammer mill.

The material is fed uniformly to the breakage zone by way of an inclined plate. The hammers are attached to the shaft so that they are free to move at right angles to the shaft through 180 degrees. They deliver heavy blows, shattering the rock and throwing it against the breaker plate. Broken pieces rebound, and the procedure is repeated until the fragments leave the mill through a grate in the bottom of the machine. The capacity of a hammer mill is more dependent on the nature of the material than is the case for most other crushers. It also relies on the velocity of the hammers and the aperture of the grate. The energy in the hammers is maintained by the kinetic energy in the rotor shaft, often supplemented by heavy flywheels attached to the ends of the shaft. Hammer mills tend to produce a cubical product because the impact action is well suited for breaking flat or elongated rocks that do not escape easily through the grate bars in the bottom of the crusher. Consequently, they are used at times for producing aggregate with a cubical shape even though high wear may occur because of hard rock.

The reduction ratio in a hammer mill is very high, up to 20:1 in open circuit and much more in closed circuit. The potentially high wear rates of the hammers and side plates generally limit the use of hammer mills to the crushing of soft, nonabrasive rocks such as limestone, coal or gypsum, or for the grinding of hay and grains for chopped feed. A rule of thumb has been that hammer mills are suitable for materials containing not more than 5% silica, doubtful for a silica content of 10%–15%, dangerous if 20%–25%, and prohibitive if it exceeds 30%.

Vertical-Shaft Impact Crushers

In hammer mills the impact of a fast-moving hammer on a slow-moving rock causes shattering. It can also be caused by the collision of a rock traveling at high velocity with another rock or a breaker plate. This is the concept of a vertical-shaft impact crusher.

FIGURE 5.19 Sectional view of the hammer mill (McGrew 1950; reprinted by permission from Primedia)

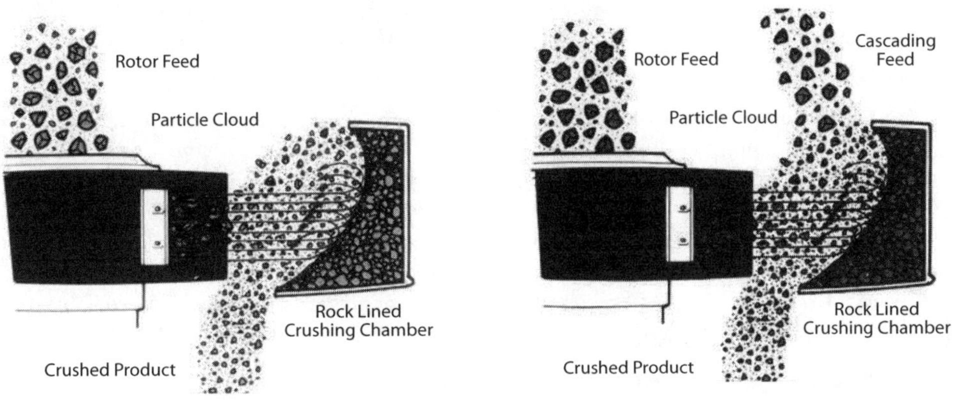

FIGURE 5.20 Crushing action in the (a) original Barmac crusher (Rotopactor) and (b) modified crusher (Duopactor), which has two feed streams (courtesy of Metso Minerals)

A problem with the original machine was that the maximum size of feed to the rotor was too small; it was 60 mm and it needed to be 100 mm. The solution was to add the larger particles as a separate flow directly into the breaking chamber. This machine was called a dual-feed machine or a Barmac Duopactor. The Barmac crusher (see Figure 5.20) was invented in 1970 by Jim Macdonald, an engineer with the Wellington City Council in New Zealand. He took the view that two principles were involved in crushing:

- Stones will break if you bang them together hard enough (thus saving wear on metal parts).
- Steel will be protected from abrasive wear if covered by a layer of trapped stones (Bartley 1996).

A basic prototype proved that the idea was sound, so, with the assistance of partner Bryan Bartley, a commercially viable machine was made. By 2002, more than 3,300 Barmac crushers had been sold, the largest being a 650-kW unit for use in an iron ore mine.

Coal Crushers

The main uses of coal are as fuel in furnaces and as a chemical to reduce iron oxides to iron. To meet environmental standards for these purposes, coal needs to be mined, crushed, separated by size, upgraded, and ground to burn efficiently with the gaseous and solid wastes. Ground limestone is also required to remove hydrogen sulfide from the stack gases from the boiler. Shale and minerals are separated from the coal during breakage and can be rejected, particularly pyrite, so that a cleaner grade of coal can be burned in the boilers. Coal crushers changed little during the 20th century, and three types of machines are still in common use:

- Conventional high- or low-speed crushers to break run-of-mine coal to pass a size that may be up to 150 mm, depending on the next process
- Drums in which the coal is lightly broken to separate the shale and minerals from the coal macerates so that the macerates can be concentrated and make a high-quality coal for use in coke ovens
- Roller mills and ball mills in which very fine particles are produced for combustion in boilers

Slow-speed single rolls with long and short teeth rotating at about 200 rpm have been used for the coarse-crushing run-of-mine coal since early in the 20th century (see Figure 5.21). Two counter-rotating toothed rolls may be used instead of one roll and a breaker plate.

> *The machine...gives a good cubical product of any size between 6 in. and 1.25 in. with a minimum of fines. ...The segments have a combination of long and short teeth, the long feeder teeth gripping the large pieces and reducing them to a smaller size and which are finally reduced to the desired size by the short teeth. Narrow slots in the shoe end of the breaker plate provide clearance for the long feeder teeth while allowing the breaker plate to be set close up to the roll to prevent oversize. ...A 36 in. size breaker is capable of giving 500 tons of 8 in. coal or 155 tons of 1.25 in. coal per hour at 240 rpm utilizing 90 hp. (Lebeter 1949a)*

Finer products could be obtained with higher-speed machines that used more power per ton, and the hammer mill became the first widely used high-speed crusher. It is particularly suitable for the fine crushing of softer, brittle solids and continues to be a leader in this field. "In the United States, impactor and hammer mills crushing coal normally are used to prepare the relatively fine crushed sizes used for charging coke ovens, feed to fine grinders and pipeline transportation" (Leonard 1991).

Carr's disintegrator, shown in Figure 5.22, was devised at the end of the 19th century to crush coal to fine particles. This high-speed machine ran at 200 rpm, and had parallel discs rotating in opposite directions. The speed of the outer disc was up to 2,000 m/min. The discs had alternating annular rows of rods, three on one and two on the other, and material fed to the center of the machine was shattered by impact with the rods as it passed through them. Eventually, the fragments were small enough to pass through the outer ring. The machines ranged from 0.6–2.5 m in diameter, and the capacity of a 1.8-m mill using 15 kW was 15–30 tph of coal.

Slow-speed rolls produced coarse lumps that contained unliberated mineral matter and could only be roughly concentrated in jigs; high-speed crushers produced too many fines that were difficult to concentrate. Neither machine was ideal for preparing coking coal for concentration because removing mineral matter was very important.

The Bradford breaker was built in the late 1800s to break coarse lumps of mined coal for coke making (Figure 5.23). Still widely used for the purpose, it was invented by

FIGURE 5.21 Toothed roll crusher (Richards and Locke 1940)

FIGURE 5.22 Carr's disintegrator, which was used for crushing coal (Louis 1909)

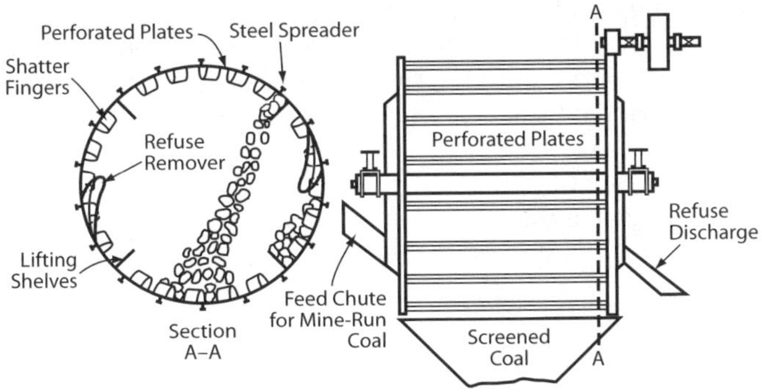

FIGURE 5.23 Bradford rotary breaker for coal (Richards and Locke 1940)

Hezekiah Bradford, but little is known about the inventor himself or the origin of the breaker.

The Bradford breaker consists of a rotary drum mounted on trunnions similar to a ball mill and rotates slowly at 12–18 rpm. The coal is raised by lifters attached to the inside wall of the cylinder of the drum and allowed to fall. The kinetic energy of the larger coal lumps is sufficient to shatter them when they hit the bottom of the drum, but the lumps of minerals and shale tend to resist breakage. The wall of the drum is made of screens, and the apertures allow particles that are small enough, usually about 50 mm, to leave the drum and proceed to further processing. The unbroken lumps of waste that do not pass through the apertures are removed through a refuse chute at the end of the drum. Abrasion breakage, which produces excess fines, is minimized, and no steel balls are used. About 80% of the macerates in typical coking coals can be broken into the size range of 1–50 mm and concentrated in dense-medium cyclones to produce high-quality coking coal.

Cryogenic Crushers

Impact breakage processes are used to reduce the size of hard, brittle materials but are not useful for reducing soft or elastic materials unless the materials can be cooled to the point at which they become brittle. This is where cryogenic breakage is becoming very important. Its main use is in recycling processes for plastic and rubber, mainly thermoplastics and scrap tires, and it works by cooling the materials below their embrittlement points so they can be shattered in an impact mill. Liquid nitrogen and occasionally carbon dioxide are used as the coolants.

One advantage of cryogenic grinding is that the process operates at a low temperature (normal dry-grinding processes often operate at temperatures above 80°C, and it's used to grind herbs and spices to reduce the loss of volatiles and to grind drugs and chemicals to reduce chemical reactions. Its main use, however, is to make rubber crumb from scrap tires.

The first patent for cryogenic breakage was granted in 1909 to Gaston Galy to increase crushing efficiency by using liquid air to cool and embrittle the material being crushed. At that time, rubber was a valued commodity, particularly for its use in the growing automobile market. But it was still comparatively scarce, and its price was so high in 1910 that 0.45 kg of rubber cost nearly as much as the same amount of silver. It is not surprising that every effort was made to recycle rubber products and that the average recycled content of all rubber products was more than 50%. By 1960, the recycled content had decreased to 20% and by 1995 to 2%. Much of the decrease resulted from the difficulty of grinding steel-belted radial tires, but recycling is increasing. In 1999, 228,800 tons, or 10% of rubber available for disposal, was recycled in the European Union and 180,000 tons, or 7%, was recycled in the United States. Tires are recycled by shredding into chips about 5 cm in size and then grinding the chips into rubber crumbs. Ambient grinding or cryogenic grinding can be used to produce the crumbs (Reschner 2002).

Ambient temperature grinding is carried out in a series of granulators, which work by a cutting and shearing motion, with the product size controlled by a screen in the mill and by cracker mills, which are counter-rotating rolls with serrations cut in them. The lower practical limit for the ambient process is 40 mesh. Cryogenic breakage requires the rubber to be cooled to less than −600°C, at which point it becomes as brittle as glass and is shattered readily in a hammer mill. Figure 5.24 shows a typical cryogenic grinding circuit. Cryogenic grinding also liberates the fiber and steel more completely from the rubber than ambient grinding, so the final product is cleaner. The temperature of liquid nitrogen is −2,000°C at atmospheric pressure, and the choice of temperature depends on the crumb size required. The consumption of liquid nitrogen is 0.5 lb/lb of rubber at −1,000°C and 0.34 kg at −2,000°C. Table 5.2 gives the sizing distributions of typical 60-mesh ground rubber from ambient and cryogenic processes.

CONCLUSION

The 20th century ended with the mining and stone industries using the same types of primary crushers that were invented by Blake and Gates and used at the end of the 19th

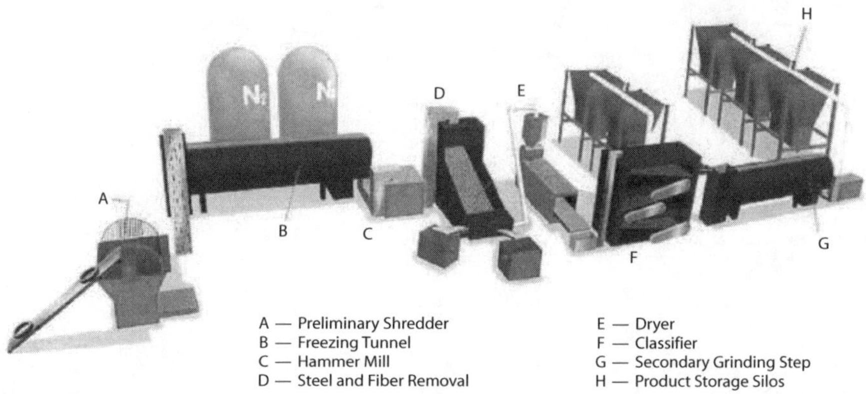

FIGURE 5.24 Typical size-reduction circuit for a cryogenic scrap tire–processing plant (Reschner 2002; reprinted by permission from Recovery Technologies Ltd.)

TABLE 5.2 Sizing distributions

Mesh	Ambient, % Retained	Cryogenic, % Retained
30	2	2
40	15	10–12
60	60–75	35–40
80	15	35–40
100	5	20

century. Crushers have become larger and can crush much larger rocks, and new designs have improved them, but the same basic crushing mechanisms are used. Cone crushers are the principal crushers used for secondary, tertiary, and fine-reduction crushing. The hydraulic pistons for tramp iron release and hydraulic main shaft support systems were used as the basis for remote systems to control the size of the discharge openings. There was considerable emphasis on redesigning existing crushers so that they could draw more power and increase productivity. A recent development in crushing is the demand for portable plants to be located near construction sites for new highways and for rebuilding existing highways.

In this chapter we covered developments in larger machines that applied more than manual energy for coarse size reduction. Waterpower replaced manual power, then steam engines replaced waterpower, then steam and water were used to drive turbines that drive generators that generate the electricity for electric motors.

Stamps and jaw crushers crush the feed and discharge the crushed material, at which point new feed is added. This, however, gives an interrupted flow of material. Because the crushing chambers of gyratory and cone crushers are conical in shape and the crushing surface swings in a gyratory motion, these crushers have a continuous feeding and discharging flow. Roll crushers have two rotating drums that impart the crushing action so are fed and discharged in a continuous flow.

The chronology of the developments in this chapter illuminates the use of available technology and energy sources to develop machinery to meet an immediate demand for size reduction. Many of the crushers discussed were used to prepare feed for the grinding mills that were being developed concurrently. In the next chapter, we discuss the development of the roller-type grinding machines that followed the fundamental principles of the quern.

. . . .

CHAPTER 6

Roller Mills

Roller mills were used in China in the 2nd century for grinding the red mineral cinnabar to a vermillion pigment (Figure 6.1) and for grinding grains for cereal (Figure 6.2), and they are still used for these purposes today (Figure 6.3). In 1449, Pietro Speciale was credited with the development of a three-roller mill in Sicily; it was originally used for crushing sugar canes, and the rolls were made of wood mounted with vertical axes. In 1558 in Spain, Giovanni Turriano developed a roller mill with a conical roller with spiral grooves cut into it that rotated inside a grooved cone (Ramelli 1588; see Figure 6.4). It was powered by hand and apparently was small enough to fit inside the sleeve of a monk's robe. Although it was said to be extremely productive, the concept was so far ahead of practice at the time that the mill was not widely used. In 1753, Issac Wilkinson in Italy obtained a patent for a mill with cast iron rolls (Research Association of the British Paint, Colour and Varnish Manufacturers 1953).

ARRASTRAS

> *In its simplest form the arrastra consists of a shallow circular bed, which is paved with closely set blocks of hard stone; upon this pavement two or more heavy drags of stone are slowly pulled round, thus crushing the ore under them. (Louis 1894)*

> *(Arrastra) drag stones weigh from 80 to 2000 lb each. ...they are placed so that the front edge of the stone may be lifted so as to ride over the coarsest of the ore during the early stage of grinding. ...the speed is 4 to 18 revolutions per minute. (Richards 1908)*

Arrastras were the first industrial process devised to meet a specific mineral processing need. They came into use in Mexico starting in the late 16th century, where they were driven by donkeys. After 60 years of mining, the high-grade ores suitable for direct smelting were exhausted, and this left only ores containing less than 200 oz of silver per ton. These ores were finer grained, the deposits were huge, and the profits to be made were high, so a process was needed to grind these finer silver ores and increase the quantity of silver being sent from Mexico to the Spanish king.

Although a chemical process in which the ore was mixed with water, common salt, copper sulfate, and mercury was not new (it had been described in the first complete textbook written on metallurgy [Biringuccio 1990]), little was known about it in Spain in 1585. At that time, Bartolome de Medina, a wealthy Seville merchant, learned from a German citizen, Maestro Lorenzo, that fine grinding and amalgamation might work. In 1587, de Medina developed the patio process, in which an arrastra ground the ore and promoted a chemical reaction by slow mixing at the same time.

> *Grind the ore fine. Steep it in strong brine. Add mercury and mix thoroughly. Repeat mixing daily for several weeks. Every day take a pinch of ore mud and examine the mercury. See? It is bright and glistening. As time passes it should*

FIGURE 6.1 Grinding cinnabar with an edge mill in China in the 2nd century AD (Reprinted by permission from Cambridge University Press)

FIGURE 6.2 Grinding cereal grains with an edge mill in China (Reprinted by permission from Cambridge University Press)

FIGURE 6.3 Horse-driven communal edge mill used in a Turkish village today

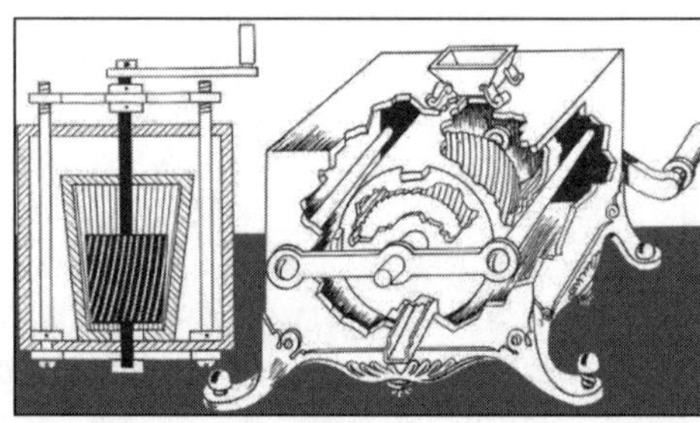

FIGURE 6.4 Original roller mill built in 1558. The internal corrugated conical section could be moved back and forth by means of the long side bolts, and this adjusted the degree of grinding (Storck and Teague 1952; reprinted by permission from University of Minnesota Press).

> *darken as silver minerals are decomposed by salt and the silver forms an alloy with mercury. Amalgam is pasty. Wash out the spent ore in water. Retort residual amalgam; mercury is driven off and silver remains.* (Probert 1969)

De Medina made the patio process work in a plant in Pachuca, and it came to be used at many mines in Mexico, leading the nation to become the world leader in the production of silver. Figure 6.5 is a lithograph depicting the patio process and arrastras used at the Hacienda de Salgado in the 1820s.

The arrastra was based on the idea of breaking particles by trapping them between a moving weight and a fixed surface, and this idea was used in several machines in the

FIGURE 6.5 Arrastras at the Hacienda de Salgado in Mexico in the 1820s (H.G. Ward, *Mexico en 1827*, London: H. Colburn [1827]; reprinted by permission from University San Luis Potosi)

1880s. The pit was 0.3–0.6 m deep and 3–6 m in diameter, and 0.3–0.4 tons was a common weight for a stone. Eventually, vertical circular stones with wide rims called tahonas replaced the drag stones in most arrastras, and in this form they became conventional edge mills. Tahonas rotated in circular tracks and could be driven easily by mules. They were typically made of basalt, with a diameter of 1.4–1.8 m and a thickness of 0.6 m, and they rotated on a flat stone, about 2 m in diameter, in a vessel shaped like a cup to prevent spilling of the ore. One or two tahonas were used in each arrastra. In many circuits, the ore was ground to −6 mm in a stamp mill, then to −0.25 mm in an arrastra after which the amalgamation process would start. Figure 6.6 shows a monument representing an arrastra with a tahona that was built to commemorate their importance to the Mexican economy.

A descendant of the arrastra is the vertical roller mill, which is used extensively for dry grinding high volumes of soft, less abrasive materials such as coal and limestone. As better abrasion-resistant materials became available, vertical roller mills were increasingly used to grind cement clinker and mildly abrasive mineral ores.

EDGE ROLLER MILLS

In the edge mill, which had been developed in China in about 200 AD, a stone or concrete disk rotated on a stone base. The ancient edge mill is still used in many communities:

> *In 1975 people in Antakya dried and ground their wheat in September for winter. This was done in meydans (village squares). An old mill was in the center of the meydan in my ward, its wheel was made from one piece of granite and was about 2 m diameter and 0.5 m wide, and the grinding surface on which the wheel rotated was marble. (Ekmecki 2003)*

Edge mills will probably always be used for grinding grains in remote villages where electricity is expensive and water wheels do not exist.

FIGURE 6.6 Monument to an arrastra with a tahona

During the 19th century the stone rollers used in the ancient edge mills were replaced by iron or steel rollers. This form of edge roller mill, which is shown in Figure 6.7, became known as the "Chilean mill." It could be operated at low or high speeds (see Table 6.1).

By 1900, the demand for grinding coal, cement clinker, and fertilizers in dry mills was increasing, and this presented an opportunity to the owners of small engineering workshops to build better mills. In 1906, two owners, Curt von Grueber in Berlin and Claudius Peters in Hamburg, visited the United States to look at grinding mills. They both returned to Germany with licenses to build mills—the Maxecon mill by von Grueber and the Fuller mill by Peters. Both men had the vision and tenacity to convert their ideas into large mill-manufacturing companies. It is interesting to note that several of today's large corporations that make grinding mills, such as Loesche, Alpine, Polysius, Pfeiffer, and F.L. Schmidt, began in the same way late in the 19th and early in the 20th centuries. Edge roller mills built in the late 19th century were known by the names of their inventors or manufacturers.

Compression and abrasion were the breakage mechanisms in the roller mills that were built in the 19th century to grind cereal grains, softer minerals such as limestone and coal, and harder minerals such as siliceous ores. These mills now grind thousands of millions of tons of solids annually.

The Huntington Mill

Figure 6.8 shows a ring-roller mill invented by F.E. Huntington in 1883 to grind gold-bearing quartz ores to liberate the gold before amalgamating it on tables. The Huntington mill was a roller mill similar to an arrastra, but centrifugal force, instead of gravity, drove its grinding action. It had three suspended flat rollers that swung radially outward against a horizontal steel die as the drive wheel rotated and ground the ore that was fed into the drive wheel's path. Pulverized ore escaped from the mill through screens placed above the level of the viscous pulp and flowed by a launder to the copper amalgamating tables. The 1.5-m Huntington mill ground up to 20 tons in 24 hours through a 30-mesh (600-μm) screen. It rotated at 70 rpm and used up to 8 kW. The power required by a 2-m mill crushing 75 tpd of −6 mm jig tailings to −1 mm was 10 kW. Centrifugal pendulum mills were soon used for clinker grinding. The Atlas Portland Cement Company reported that a 30-kW Huntington mill reduced 2.5 tph of 12–25 mm cement clinker to 92% passing 150 μm (Eckel 1905).

FIGURE 6.7 Gruson vertical roller mills (Chilean mills) in 1890 (Foster 1894)

TABLE 6.1 Characteristics of Chilean mills

	Roller Weight, tons	rpm	Feed Size, mm	Feed Rate, tpd
Low speeds	2–4	8–16	–30	20–30
High speeds	1	30–40	–6	60–100

The Griffin mill, also shown in Figure 6.8, used a similar principle but had a single roller driven directly at its upper end. This mill was used in the late 19th century for grinding portland-cement clinker.

In 1989, Horst Brundiek wrote the history of dry roller mills, which are used in the cement and coal industries. He worked with Loesche GmbH for 40 years and patented many inventions that used roller mills. Extensive use has been made of that history in writing this section, and we acknowledge his paper entitled "The Roller Grinding Mill–Its History and Current Situation" as the source of much of our information.

> *[In the cement industry] the roller grinding mill is a machine in which the grinding surface is annular. Grinding bodies [rollers or balls] roll on it. The grinding bodies are pressed down on the grinding surface either by their own weight, or by centrifugal force, by springs, or by hydraulic or pneumatic systems. Both the driving surface and the grinding bodies may be driven. (Brundiek 1989)*

> *[In the ore industry] the principle of the roll crusher is that twin rolls, not in contact, revolve so that the upper surfaces move toward each other. Friction between the rolls and fragments inserted between them will be sufficient to draw in the fragments and crush them till they pass the pre-determined gap. (Lebeter 1949b)*

Rollers have been operated in many ways in dry roller mills. The different methods of operation were defined by G. Tarjan (1981) and are shown in Figure 6.9. Three roller mills–the Maxecon ring-roller, the Raymond centrifugal ring-roller, and the Fuller ring-ball–were used in the United States at the start of the 20th century for grinding softer rocks such as phosphates and coal (see Figure 6.10).

The Maxecon Mill

The Maxecon mill had a vertical grinding ring rotating around a horizontal axis and three convex horizontal rollers that rotated independently and were pressed against its

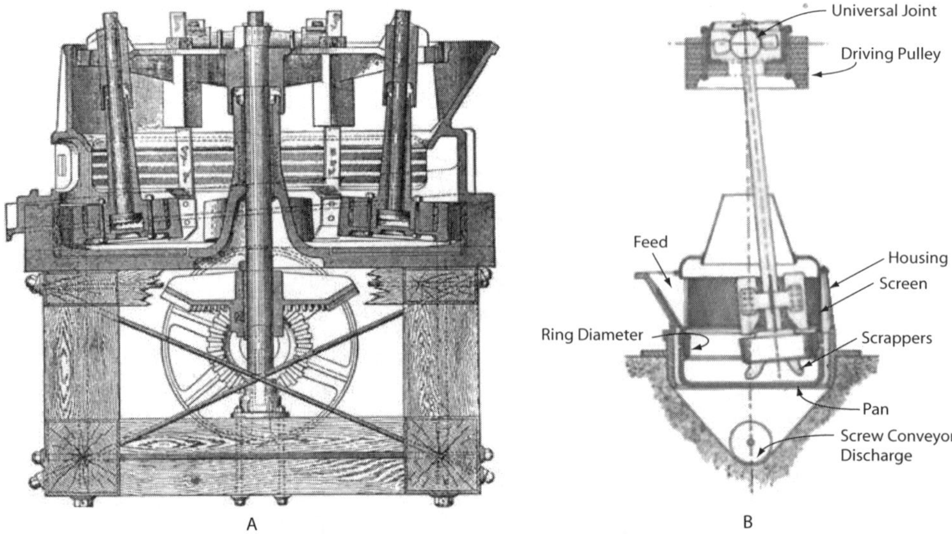

FIGURE 6.8 (a) Huntington ring-roller mill showing the rollers attached to vertical axles, the discharge screen just above the rollers, and the fixed ring against which the rollers did the grinding (Louis 1909) (b) Griffin single-roller mill (Truscott 1923)

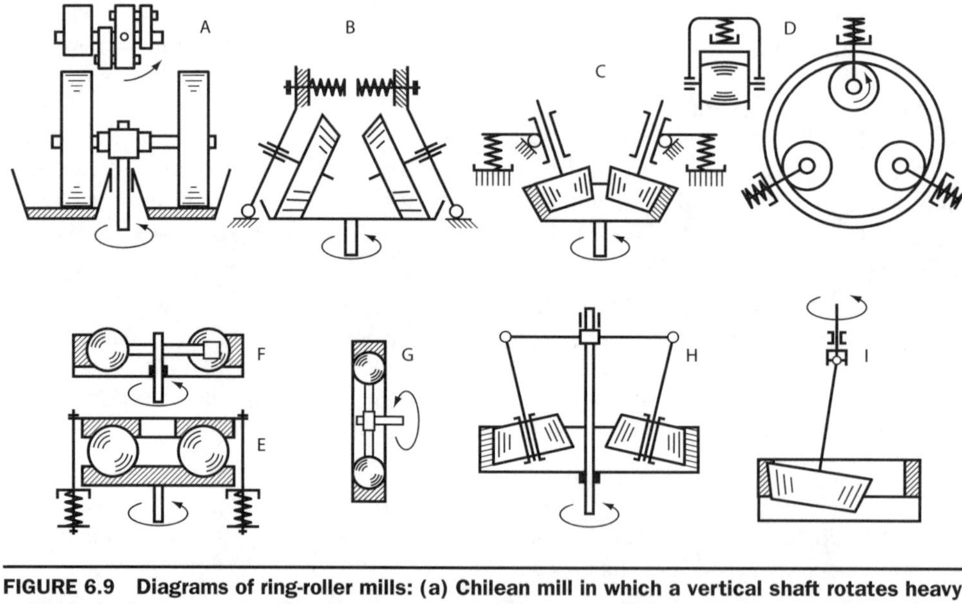

FIGURE 6.9 Diagrams of ring-roller mills: (a) Chilean mill in which a vertical shaft rotates heavy cylindrical rollers around a plane ring race (b and c) Loesche mills and disk-type Raymond mills in which rollers are pressed by springs against the race that rotates (d) Maxecon mills and Kent mills in which three convex rollers are pressed by springs against a complementary concave race fixed to a rotating ring (e) Babcock & Wilcox mill or Fuller-Peters mill in which roller balls are placed between two races, the lower being driven and the upper being pressed by springs against the balls (f and g) Fuller mills and Sturtevant roulette mills in which spherical rollers driven around a stationary ring are pressed by centrifugal force against the race (h and i) Raymond or Bradley multi-pendulum mill and Griffin single pendulum mill in which the grinding pressure results from centrifugal force as the rollers attached to each arm suspended from a pivot press against a fixed race as the shaft rotates (Tarjan 1981)

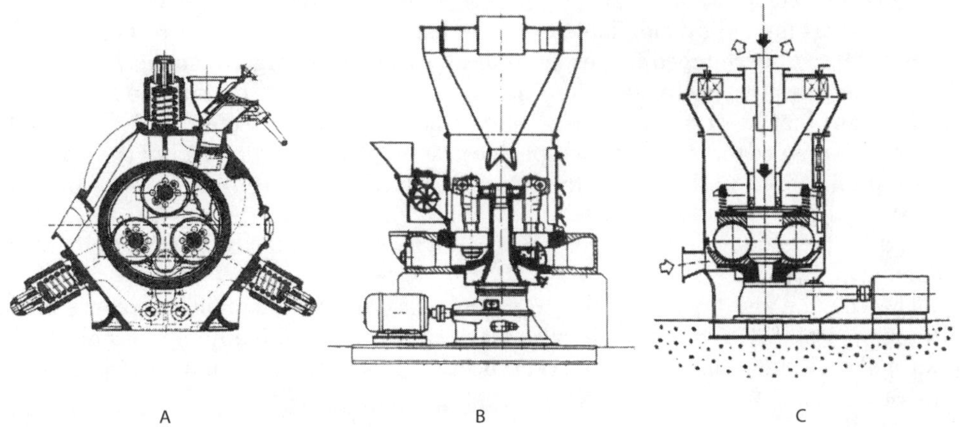

FIGURE 6.10 (a) Maxecon ring-roller (b) Raymond centrifugal ring-roller (c) Fuller ring-ball mills (Brundiek 1989; reprinted by permission from Loesche)

concave inner surface. One roller was driven by an external motor and moved the grinding ring by friction. The feed entered the grinding zone, and centrifugal force held the particles at the ring and as they passed under the rolls. The product was discharged from the edge of the ring. The Maxecon mill was very successful, and many were sold by von Grueber's company for grinding phosphate rock and limestone. Eventually 600 machines were supplied to various industries. In 1924, two Maxecon mills were used as coal pulverizers at the Moabit power station operated by Berliner Elektrizitatswerke (BEWAG), and an indication of their success was that two more were added when the station was expanded. Their capacity was 5 tph, but the possibility for increasing the capacity was limited because their geometry permitted only a small increase in roller diameter.

The Raymond Mill

The Raymond mill had rollers attached to vertical shafts that were suspended from a horizontal beam through joints that allowed them to swing. Rotation of the beam caused the rollers to swing outward and press against the grinding ring. The feed was introduced into the grinding zone between the rollers and the ring, and the product was extracted by air that was pumped into the grinding chamber through grates. Raymond mills worked well mechanically but were less successful at grinding the hard German coal than the soft American coal, because higher centrifugal forces were required and the vibrations damaged the mill foundations. Their proven performance in the United States on the softer minerals ensured that Loesche found a market for them in grinding limestone and phosphate rocks, but the original design had to be changed to work efficiently on coal at higher capacities.

The Fuller Mill

The Fuller mill consisted of a U-shaped grinding ring with large steel balls rotating inside that were pressed down by a spring-loaded ring. The feed entered the grinding zone from the center of the ring and passed under the balls. The ground product was discharged from the other side of the ring.

By 1925, Ernst Curt Loesche had become part owner of von Grueber's company, and he was given the contract to build mills for the new Klingenberg power station in Berlin that ground 10 to 12 tph. He had already obtained the license for Raymond ring-roller mills, which were being made in the United States, and he used these mills for the project.

Claudius Peters had a similar experience when he found that the capacity of the Fuller mill was limited by the diameter of the balls, so after his license to manufacture the mill expired, he changed its design to use fewer balls of larger diameter. He gained another increase in capacity by increasing the diameters of the housing and the bowl, which meant that he was able to use more grinding balls. The Peters mill was mainly used for grinding coal that had a low specific gravity. The large diameter of the housing, needed for the large bowl, reduced the velocity of the air that was used to lift the product from the mill to the classifier. Materials with a higher specific gravity were difficult to suspend in the low-velocity air. Low specific gravity coal was the most suitable feed for the Peters mill.

The Loesche and Peters mills were roller mills, which were similar in concept but different in design. Other companies also made similar mills with different designs. Each roller mill took some share of the market, because it was suitable for a specific application but

> *...for the most part the roller mills available on the market differ only in the comminution zone. There are differences in the geometric form of the grinding track and grinding elements, in the arrangement and number of grinding elements, and in the way they are held against the grinding bed. (Feige 1993)*

Figure 6.11 summarizes four types of configurations of grinding tracks and grinding elements that have been widely used in roller mills.

In the United States, a ring-ball mill that was similar to the Peters mill was built by Babcock & Wilcox and first used in 1929 at Commonwealth Edison's Powerton Station in the New York City metropolitan area. Known as the Babcock E mill, it contained a larger quantity of smaller balls than the Peters mill. These mills were used in England for pulverizing coal for boilers. About the same time, a horizontal three-roller mill with a screen discharge similar to the Griffin mill was built by the Bradley Hercules Company and used in a roller mill–tube mill circuit to grind cement clinker. The power consumed in grinding was low compared with two-stage or two-compartment ball mills, but the wear on the screens and side plates was severe, and there were high production losses resulting from the time required to repair the mills. At least one standby mill was required for every two operating mills installed.

Rod mills were also tested for grinding cement clinker. They consumed about the same amount of energy as a Bradley Hercules mill but placing rods in the mills required 1 hour or more every 2 days, because they had to be added one at a time through a small hole in the dust cover on the discharge end trunnion. This was not acceptable in cement plants.

INNOVATIONS IN ROLLER MILLS

Many roller mills were sold from 1930 to 1950, but there was little enthusiasm for new designs because of the economic depression and World War II. The revival of industry after the war increased the demand for materials, and new mills were required with grinding chambers that had higher processing capacities.

Fixed Rollers and a Spinning Bowl

By 1925, Loesche's experience with the Maxecon and Raymond mills persuaded him that fixed rollers and a spinning bowl was better than spinning rollers and a fixed bowl, which was the mechanism in the Raymond mill. With this reverse design the rollers could be spring-loaded and higher grinding forces could be applied than those generated by the centrifugal forces in the Raymond mill. With the fixed roller–spinning bowl system

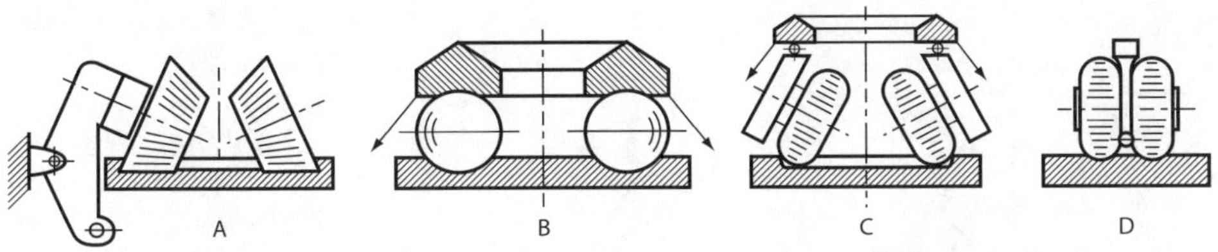

FIGURE 6.11 (a) Loesche mill, tapered-lever mounted rollers (b) Peters ring-ball mill, track-guided balls (c) Pfeiffer mill, track-guided crowned rollers (d) Polysius roller mill, track-guided crowned rollers (adapted from Feige 1993)

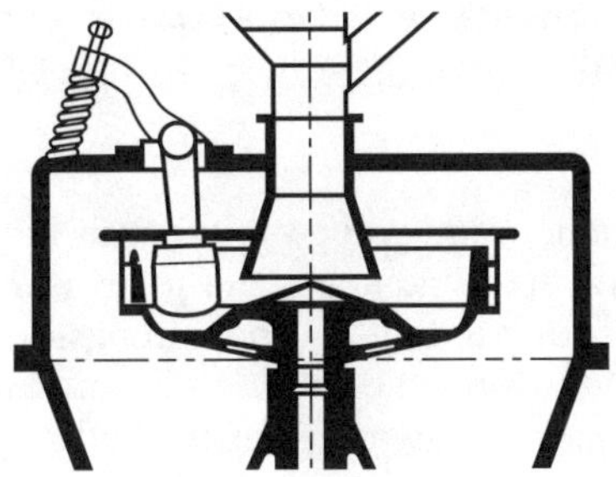

FIGURE 6.12 Maximal mill with spinning bowl and fixed rollers (Brundiek 1989; reprinted by permission from Loesche)

the material was fed into the center of the bowl, flowed toward the wall by centrifugal action, and ground by passing between the rollers and the bowl wall. Loesche's first mill using a spinning bowl was named the Maximal mill (see Figure 6.12).

There was still the problem of geometry. The rollers rotated in a horizontal plane and had to fit inside the wall of the bowl, and this gave little scope to increase the roller diameter and mill capacity. The solution was to incline the wall of the bowl backward so that the roller axles were at a 45-degree angle that removed the restriction on roll diameters. The angle of inclination of the axles was reduced in stages and reached 15 degrees by 1935, which is the current value for Loesche mills. The first of these mills, built in 1928, had backward-sloping bowl walls and the rollers were held in rocker arms. The new mill was offered to BEWAG, which accepted the machine on condition that it would be called the Loesche mill. This condition was intended to make it absolutely clear that if the mill failed, the inventor of the mill carrying his name was responsible for the functioning of the machine and not the BEWAG department dealing with new construction (Brundiek 1989). In the early air-swept roller mills, efficient classification was necessary, and static centrifugal classifiers were used but were soon replaced by dynamic classifiers to give better control of particle size. By 1937, 400 mills had been sold, and the mills remained as popular as the difficult circumstances permitted from 1937 to 1952. The largest Loesche mill available during this time was 2 m in diameter, and the grinding elements were two 1.5-m diameter rollers. It was used for grinding cement raw materials at 50 tph.

Early in the 1960s, dry grinding replaced wet grinding ahead of the cement kiln, and roller mills were used for grinding limestone. The largest mill available then could grind 120 tph of limestone–large enough to produce 1,500 tpd of cement clinker, which was

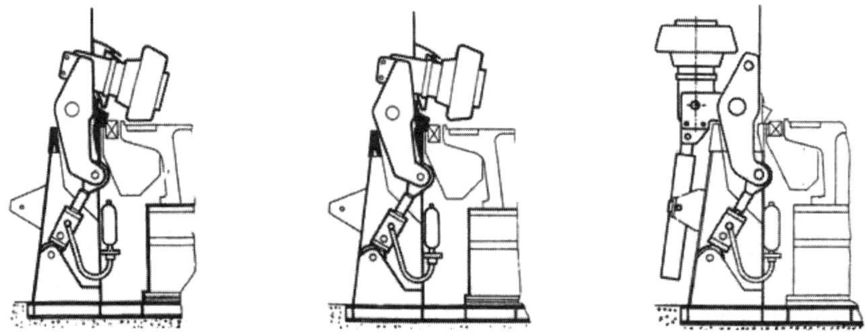

FIGURE 6.13 Positions of roller mounted on pedestal in Loesche roller mill: (a) operating position (b) start (clear of bowl) (c) service (swung away from bowl) (Brundiek 1989; reprinted by permission from Loesche)

the maximum capacity of cement kilns. By 1970, better kiln technology had increased kiln capacity to 3,000 tpd, and mills had to be built to grind 240 tph of limestone. Making larger rollers was not a problem, but holding them in position against the spinning bowl while they were grinding and raising and lowering them easily at other times was more difficult. To do this, a module system was designed that comprised a roller, rocker arm, and spring-loading system mounted on an individual pedestal. The operation of this module is shown in Figure 6.13.

This development opened the door to designing very large roller mills, because huge rollers could be installed in a machine and handled easily in operation. In 1999, a 750-tph mill that had four wide rollers 2.7 m in diameter by 0.9 m long was installed at the Siam Cement Company's Thung Song plant in Thailand. This mill was driven by a 5,400-kW motor (Ongtkoon 2000; see Figure 6.14).

There are always many engineering problems involved in building and assembling very large grinding mills; for example, the grinding table in the Thung Song mill weighed 230 tons and had to be cast in one piece.

The advantage of roller mills was that they required less energy per ton than ball mills to grind limestone to the same size. Clinker grinding is the high-energy consumer in cement plants, and this pointed toward using a roller mill for grinding cement clinker to finished cement. The problem was that a bed of cement particles on the grinding table was unstable, because (1) it was very fine, containing up to 10 times the volume of classifier reject as of new feed, and (2) it contained occluded air.

Aerated fine particles flow quickly across the table and are not ground efficiently. The method adopted to improve the stability of the bed was to add a small roller ("S-roller") ahead of each main roller ("M-roller"). The S-rollers were set well above the bed and prepared it for the M-rollers by consolidating and partly de-aerating the bed (see Figure 6.15). These mills are now used for producing finished cement.

Track-Guided Rollers

The MB mill was designed and built in 1947 by Max Berz, formerly head of the design department at von Grueber's company. It had three large rollers that rotated in a tracking groove and were pressed down by a thrust ring. The rollers were not fixed by roller bearings and could move relative to each other when they were driven by the rotating grinding track. This limited the speed and capacity of the mill and was not very successful.

Gebruder Pfeiffer AG licensed the mill in Germany. This company grew out of a small workshop that was established by Gebruder Pfeiffer in Kaiserslurten in 1864 and

FIGURE 6.14 The 6.3-m diameter table and 2.7-m diameter rollers of a Loesche mill during installation at the Siam Cement Company's Thung Song plant in Thailand. The four pedestals are at different stages of construction (Ongtkoon 2000; reprinted by permission from Loesche)

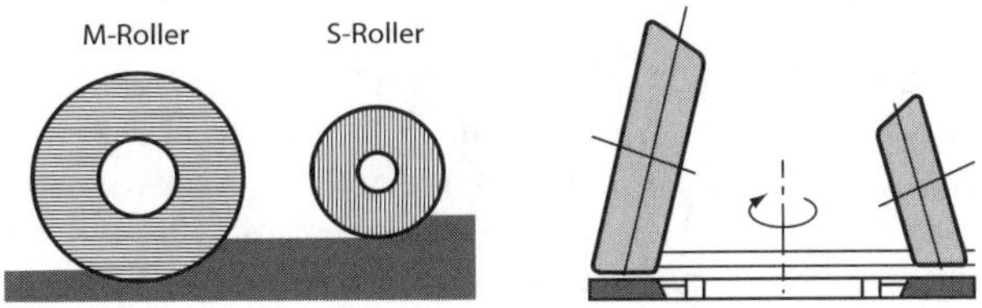

FIGURE 6.15 Rollers in series preparing and grinding the bed of particles (Schaefer 1999; reprinted by permission from Loesche)

which specialized in designing and making equipment for cement plants. Engineers soon changed the design of the MB mill to locate the rollers in fixed positions, and the mill was then called the MPS mill. They used a thrust frame to supply the grinding force to the rollers, and this permitted the rollers to realign themselves with the track as they wore. In the United States, Allis-Chalmers began manufacturing dry-grinding systems in 1970 using Pfeiffer rolls under a license agreement with Gebruder Pfeiffer AG. The Pfeiffer rolls were found to use almost 50% less power than ball mills.

The Polysius Company introduced a roller mill in the 1960s to add to its range of equipment for manufacturing cement (see Figure 6.16). It consisted of two sets of double rollers running in tracks on a rotating grinding table with each double roller suspended in a carrier. The grinding force was applied to each roller pair by hydraulic cylinders. The material being ground entered the mill at the center of the table and passed across the two tracks before being discharged. Rollers were interchangeable, and vertical movement allowed automatic compensation for wear.

In 1970, Babcock & Wilcox introduced the MPS mill to the U.S. market for pulverizing coal in power stations, and more than 1,000 mills with motors up to 930 kW (1,250 hp) and capacities up to 105 tph were sold within 20 years. The company had been building ring and ball mills for pulverizing coal since the late 1920s and by 1990 had installed more than 1,600 mills for industrial uses.

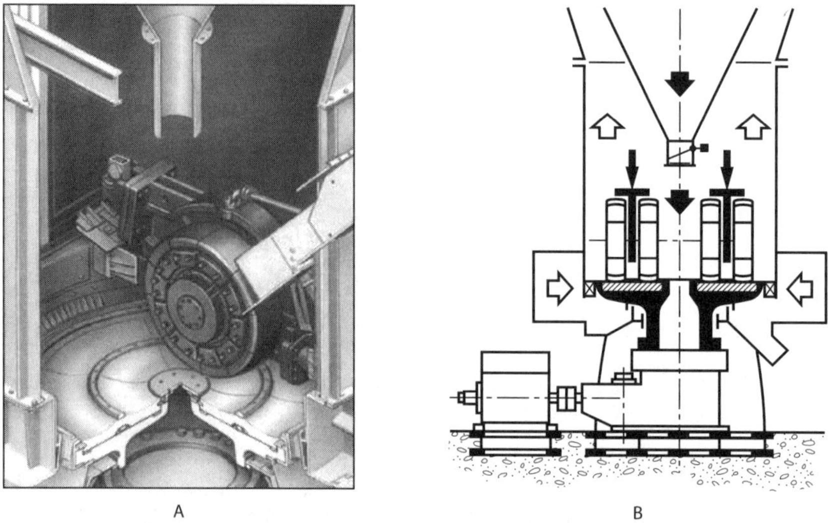

FIGURE 6.16 Polysius vertical roller mill: (a) inside the mill (Courtesy of Krupp Polysius, a Company of ThyssenKrupp Technologies) (b) mechanism (Brundiek 1989; reprinted by permission from Loesche)

The predominate use for roller mills has been for the dry grinding of coal, cement raw materials, and nonabrasive industrial minerals for the chemical industries. When roller mills were used for coarse and fine grinding of cement clinker, they used less power than ball mills; however, the abrasive characteristics of clinker are much greater than those of limestone. Major shutdowns of roller mills are required to replace the wearing surfaces, and these shutdowns are frequent with roller mills that are used to grind cement clinker. Grinding balls in a ball mill or in a multicompartment ball mill is the major source of steel consumption in these mills. Adding grinding balls to a ball mill does not require a time-consuming and costly shutdown; balls can be charged to a single-compartment ball mill without shutting it down. This is a good reason not to use roller mills to grind cement clinker.

Classifiers

In conventional dry roller mills, air is used to transport the particles from the grinding table to the classifier inlet and through the classifier, which is built into the mill housing. The fan and air transport system can take as much power as the mill, and this is shown in the power data (see Table 6.2) given for roller mills that grind limestone, coal, and clinker in the Morning Star Cement Hon Chong plant in Vietnam—the first plant designed to operate with only roller mills for grinding.

Very high airflows are needed to lift the coarser particles from the grinding table discharge to the separator, which is high above the table. This is an expensive process. Some mills are designed with an airflow that is too small to entrain the coarser particles, and these fall to the base of the machine from which they are lifted by an elevator to the separator. There is some indication that considerable savings in energy can be made by using a mechanical elevator to lift the particles rather than air blowing (Feige 1993), and this is being built into some dry-roller mill systems. Roller mills are now rated at 840 tph and can grind cement raw materials with feed moistures up to 25%. Table 6.3 shows how the throughput ratings of Loesche mills have increased over 70 years.

TABLE 6.2 Power data for roller mills

Mill	Mill Motor, kW	Fan Motor, kW	Total, kW
Raw	2,500	2,700	5,200
Coal	500	500	1,000
Clinker	2,800	1,380	4,180

Source: Fisch 2000.

TABLE 6.3 Growth in throughput ratings of Loesche roller mills

Year	1930	1940	1950	1960	1970	1980	1990	2000
Rating, tph	20	40	40	50	210	420	520	840

Source: ZKG International 2000.

The comment by Brundiek (1989) that "The outlook for further use of roller grinding mills is encouraging from the technological aspect and extremely interesting from the economic aspect" points the way to the future.

High-Pressure Double-Roll Compactors

In 1848, William Easby was granted a patent for a method of converting fine coal into solid lumps. In his application, Easby made only one claim: "The formation of small particles of any variety of coal into solid lumps by pressure." In an equally brief description of the process, he mentions, "The utility and advantage of the discovery are that by this process an article of small value and almost worthless can be converted into a valuable article of fuel for steamers, forges, culinary and other purposes thus saving what is now lost." In his few words, Easby patented the entire coal briquetting industry and also stated the rationale for its existence. Almost 50 years later, economic pressure joined forces with technological progress to give substance to Easby's vision (K.R. Komarek Inc. 2002).

Easby added high-pressure machines to the armory of equipment that could be used to add value to solid particles, and, although they took a long time to make their mark, they were eventually proved to be very useful for both compacting and disintegrating solids. The first use of the high-pressure rolls was for compacting coal into briquettes, and by 1900 coal briquetting was a large industry in Europe and the United States. The many huge buildings housing the presses reflected the high volume of briquettes produced. Figure 6.17 shows a briquetting press in the Wachtberg lignite briquette plant in Frechen, Germany, which opened in 1901. Today's presses work at 2,000 bar (29 psi).

High-pressure rolls were also built for duties other than forming lignite briquettes:

> *During the 1920s Allis-Chalmers started to build double-roll machines to compress fine solids into granules and flakes and with the development of hydraulic pistons the company was able to use the basic design of their granulators and flaking mills to manufacture a double-roll compactor. The fine material to be compacted is fed continuously into the nip of the rolls from above. This material is drawn between the rotating rolls where very high pressures are developed which would compact and agglutinate the feed so that a continuous sheet of product was ejected from the bottom of the rolls. The void content of the product may approach zero and the sheet thickness may vary from 0.25 to 0.020 inches. (Kurtz and Barduhn 1960)*

Figure 6.18 is a flow sheet of the circuit in which the machine was used.

A B

FIGURE 6.17 Wachtberg briquette plant in Frechen, Germany, which commenced operations in 1901 and in which high-pressure rolls were used to compact lignite into briquettes: (a) press house (b) machinery (Harald Finster 2002; reprinted by permission)

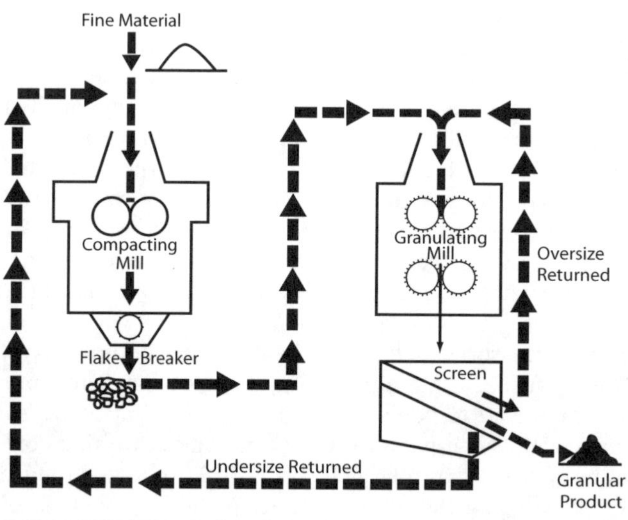

FIGURE 6.18 Flow sheet of the circuit in which a compactor was used (Kurtz and Barduhn 1960)

During their passage through the gap between the rolls, the feed particles reoriented to reduce the void content. Next, the larger particles were crushed when the minimum void was approached, and this was followed by plastic deformation to minimize the volume.

The idea that a roll mill operating at very high pressure could be used as a crusher rather than a compactor was the basis of a mill—described in the next section—that was developed 70 years later in Germany.

High-Pressure Grinding Rolls

The high-compression roller mills used in cement plants in the 1980s were similar in principle to the double-roll compactors built in the 1850s and 1950s, although apparently the only compactor used as a high-compression mill at that time was a secondhand unit installed by the Fuller Company in its laboratory as a test unit for grinding clinker. The written discussions of high-compression mills describe both comminution and agglomeration, although there is a different emphasis on each mechanism according to the purpose of the machine. High-compression roller mills became known as high-pressure grinding rolls or HPGRs.

When used for comminution, the main application for high-pressure rolls to date has been for clinker grinding in the cement industry. Cement clinker contains lumps up to 50 mm, and finished cement must be less than 90 µm. The ball mills traditionally used for producing finished cement are energy intensive. Other mills such as ring-roller mills, hammer mills, and dry-grinding rod mills were used without success as primary mills; the dry rod mills were the most efficient, but the difficulties associated with frequent rod charging made them unacceptable (Rowland and Nealy 1969). The question was: Could a mill be devised that could be used in a grinding circuit in which less energy would be used per ton of finished cement?

During the 1970s, Klaus Schonert studied the breakage of rocks and other solids when subjected to very high pressure. In 1982, he was awarded a U.S. patent on a high-pressure roll crusher for the fine and very fine comminution of brittle solids (Schonert 1982), and a few years later he published several papers in which industrial uses of HPGRs were discussed (Schonert 1988). These rolls soon became widely used in the cement industry because of their low power consumption. We believe that Schonert's patent was the most important one associated with dry grinding issued in the 20th century. At our invitation, Schonert wrote a brief review of progress from laboratory tests to the issue of the patent in 2002 (see sidebar). Several hundred HPGRs are now in use around the world.

On the matter of feed size, Schonert wrote that

> *Maximum feed sizes that can be handled by the high pressure rolls can be equated to approximately 3.5 times the resultant gap (between the rolls) for a particular material. Thus for the larger diameter rolls material with a maximum feed size of around 80 mm can easily be accepted. ...the high pressure rolls can produce in one pass a rather large amount of fines, that is 20% of −32 µm material in the case of cement clinker. (Schonert 1988)*

By 2000, high-pressure rolls were being used in many cement plants, and the energy savings in the circuits containing rolls and compartment mills were reported to be up to 50%, as compared with conventional ball mills. High-pressure rolls have also been used in ore concentrators, but the results have not always been satisfactory. They have been successful in diamond concentrators, because they appear to promote diamond liberation and the ore is relatively soft. With copper ores, however, the same wear problems that beset double-roll crushers for many years recurred, although new materials increased the wearing lives of the roller surfaces. It is likely that advances in materials technology and new designs for roller surfaces will continue to reduce the problems.

OTHER APPLICATIONS OF MULTIPLE-ROLLER MILLS

The different types of roll mills that have been used for pastes include three-, and even five-roller mills. During the latter part of the 19th century, double-roll mills were adapted for use in the grain, ink, paint, and rubber industries, and they are still used for these purposes. Each application has specific objectives; for example, grains must be torn open so that the endosperm can be separated from the husk and the husk removed as waste; pigments must be dispersed completely in the paints or inks; and rubber must be shredded. The grain industry is, by far, the largest user of grinding rolls.

Roll mills were suitable for grinding materials with very high viscosity, which could be up to 5,000 poise for pastes and higher for rubber. Rolls were used for grinding in the

THE HIGH-PRESSURE COMMINUTION PROCESS

by Klaus Schonert

Extensive research in the late 1960s on single-particle breakage in the fine size range (100 μm–3 mm) and model calculations of a multistep single-particle cascade, which can be considered as an ideal comminution process, have shown that ball milling has an efficiency of only 5%–10% (Schonert 1967). These results were proved with a lab-scale roller mill instrumented with a torque-meter and fed in such a way that the particles are broken in the roller gap without interfering with each other. The experimental procedure was designed to represent a multistep single-breakage cascade: After each run the product was sieved, and only the coarsest fraction fed again to the mill whose gap had been adjusted to a smaller width. In the last run the gap width measured 50 μm (Schonert, Ohe, and Rumpf 1965). This work showed that powders of brittle materials could be produced with very low energy expenditure by single-particle breakage. However, the throughput of a roller mill is proportional to the gap width, and a huge number of mills are needed to achieve a reasonable capacity. An array of hundreds of roller mills performing single-particle breakage would not be a solution for low-energy comminution in practice.

One point was clear—saving energy in the milling of brittle materials needs to avoid particle interference during stressing and to discharge all fragments with the fineness of the desired product immediately after being produced. To meet the first demand, we investigated with a piston press the breakage of quartz and cement clinker in a mixture of particles and small steel balls. The ball diameter was varied between some millimeters and some hundred microns and the pressure kept below 10 MPa to avoid agglomeration. The results did not show a significant effect of ball diameter on comminution efficiency. By accident, some experiments were done without steel balls and the efficiency was the same. Our expectation that adding small steel balls would cause the single-particle stressing condition to be approached in a particle bed failed.

After this disappointment we studied extensively the interparticle breakage in the particle bed. Again the pressure was kept small according to the general opinion that agglomeration has to be avoided because this worsens the comminution. From this research we learned a lot about interparticle breakage and improved our understanding of roller-table mills. The limitation with respect to agglomeration restricts the comminution effect at one stressing event; therefore, many stressing events with intermediary classification are needed to achieve a fine product. The internal circulating loads in such mills are quite high. For a better understanding, I decided to extend the pressure range up to several hundred megapascals. The briquettes produced were deglomerated by stirring in a liquid (water or methanol). These experiments provided us with the characteristics of interparticle breakage applying high pressures.

The next step was to investigate the deglomeration of the briquettes in a lab-scale ball and impact mill. With all these data we proposed an unusual comminution process consisting of high-pressure interparticle breakage followed by a deglomeration in a ball or impact mill. The astonishing results were that, in this way, only one-third to one-half of energy is needed compared to ball milling. In 1977, I applied for a patent claiming a two-step process consisting of stressing a particle bed with a high pressure above 50 MPa and a succeeding deglomeration.

High-pressure interparticle breakage can be performed in machines with different designs. With respect to the force flow, a two-roller machine is the simplest one. For this reason we continued our experimental and theoretical research on high-pressure roller mills. As far as the pressure is concerned, I want to mention the following:

- Increasing pressure decreases in principle the comminution efficiency of interparticle breakage but raises the production of ready material, by which the recirculation is reduced.
- The recirculation determines the size of mill, classifier, and transport equipments and, by that, the investment costs. Increasing pressure increases wear.
- The comminution costs result from investment, energy expenditure, and wear. An economical optimization of a high-pressure comminution has to consider all three facts.

My patent (Schonert 1982) claims the high-pressure comminution process, not the high-pressure roller mill. In my understanding, only the idea to stress a particle bed by a pressure much higher than usual, and not to be afraid about agglomeration, is the essential point of the novel comminution method. One should remember the general principle that any agglomeration worsens the comminution effect and should be avoided. Contrary to that, in a high-pressure roller mill, the material is more or less briquetted. This argument was always essential for winning the big patent cases in Germany, the United States, and Denmark.

ink, paint, and rubber industries. In 1825, a roller mill was used in England for masticating rubber (Fischer 1944), and they are still used in the rubber industry. In the paint and pigment industry roller mills are now only used for inks. They work by applying shear and heat to the resin that contains the pigment; these soften the resin and the pigment is dispersed as the sheet of resin is passed repeatedly through the rolls.

CONCLUSION

The grinding component of size reduction uses the most energy and has the highest wear on the grinding surfaces. Roller mills, whether the ring-roller type or the double horizontal roll type, use the least energy but cost the most to maintain because of the high rate of wear on the surfaces. In the next chapter, we describe the tumbling mills (including ball mills) that were developed in the last half of the 19th century and saw dynamic growth in the 20th century, becoming the most widely used mills in the world.

CHAPTER 7

Tumbling Mills

By the middle of the 19th century, the need for better fine-grinding mills had become apparent. Stamps driven by steam engines had limitations on how fine they could grind, particularly on the size of the screens used to remove the desired product from the breaking chamber and keep the oversized material in the chamber. Querns were also sometimes used for fine grinding. As discussed in Chapter 6, the roller mills (a modernized version of the quern) had high maintenance costs when used for grinding abrasive materials and also were unsuitable for wet grinding. These technologies, including the Cornish double-roll crushers, produced a sufficiently consistent product fineness for mineral liberation. At the time, a large volume of cement clinker was being produced, which had to be ground very fine to make finished cement. In addition, deposits of coarse-grained ores were depleted, and the fine-grained ores being mined required finer grinding. Inventors, then, turned their attention toward fine-grinding machines, and tumbling mills were developed.

INTRODUCTION TO TUMBLING MILLS

> *The tumbling grinding mill was not just an adaptation, it was an invention because it required thinking on a somewhat higher order–there was no prototype. (SME-AIME 1985)*

Tumbling mills use horizontal rotating cylinders that contain the grinding media and the particles to be broken. The mass moves up the wall of the cylinder as it rotates and falls back into the "toe" of the mill when the force of gravity exceeds friction and centrifugal forces. Particles are broken in the toe of the mill when caught in the collisions between the grinding media themselves and the grinding media and the mill wall. In tumbling mills the grinding media and particles acquire potential energy that becomes kinetic energy as the mass falls from the rotating shell. Tumbling mills are customarily divided into categories that are mainly defined by the size of the feed particles and the type of grinding media:

- Ball mills, which use steel or ceramic balls, are mainly used for fine grinding and are divided into two types:
  - Tube mills, which usually have a high length:diameter ratio (~6:1) and two compartments separated by a partition
  - Single-compartment mills, which have a small length:diameter ratio (~1.5:1). Single-compartment ball mills are the best-known form of tumbling mills
- Pebble mills, which use beach pebbles or small worn rocks, are fine-grinding mills.
- Rod mills, which use steel rods as grinding media that are 50–100 mm in diameter and extend the length of the mill, are fine crushers.

- Autogenous mills, which use the large rocks in the feed to break small pebbles by impact while the large rocks themselves are worn by abrasion until they are small enough to be broken by impact. The word "autogenous" means self-generating—the feed itself continuously creates the grinding media.
- Semiautogenous (SAG) mills, which are conventional autogenous mills in which 4%–12% of the mill volume is filled with large steel balls because of a lack of a sufficient quantity of grinding media in the feed. The name SAG mill is imprecise but has become accepted through common usage for primary mills, whether fully autogenous or semiautogenous.
- Secondary autogenous mills, which are mills using pebbles screened from the products from primary autogenous or SAG mills, or secondary crushers. These mills are some times referred to as pebble mills. The length:diameter ratio is similar to pebble mills, and the grinding media are pebbles extracted from the ore.

FUNDAMENTAL FACTORS COMMON TO ALL TUMBLING MILLS

Regardless of the category, all tumbling mills operate on the same principles:

- The total weight of the charge in the mill—the sum of the weight of the grinding media, the weight of the material to be ground, and any water in the mill—is a function of the percentage of the volume of the mill it occupies.
- The power the mill draws is a function of the
  - Weight of the charge in the mill
  - Percentage of volumetric loading of the mill
  - Percentage of critical speed, which is the speed in revolutions per minute at which the outer layer of the charge in the mill will centrifuge.
  - The length of the lever arm from the center of the mill drum to the center of gravity of the charge
  - The angle above horizontal at which the tumbling of the charge starts
- All the elements of the rotating drum go through a complete stress cycle from compression to tension and back to compression during each rotation of the mill.

The development of tumbling mills was heavily influenced by three industrial processes:

- The grinding of clinker for the cement industry (up to about 1900)
- The cyanide process, which was developed to remove gold from finely ground gold ore (1900–1910)
- The flotation process, which was developed to remove sulfide minerals from mineral ores that had been ground fine enough to liberate the mineral particles (1910–the present)

A discussion of these influences is necessary to put the development of tumbling mills and fine grinding into context (Coghill 1929).

THE CEMENT INDUSTRY

The use of a mixture to bind materials—called "mortar"—to hold rocks together in a building or wall was developed in the Stone Age, and, by 1811, burnt lime had been used as a cement for more than 2,000 years. In that year, Joseph Aspdin, an English mason, burned a mixture of lime and clay and obtained a better cement, which he called portland cement, because it resembled rocks found at Portland, England. It proved to be an excellent material for making concrete for building pathways, streets, and floors in domestic and commercial buildings. Aspdin obtained a patent in 1824. He used a bottle kiln with a capacity of 90 barrels to produce the cement, but the burn took several days to complete so the production rate of finished cement was very low. In 1848, his son William Aspdin built a kiln in which sintering occurred, and this was the start of the production of portland cement as it is known today.

The demand for portland cement grew rapidly, requiring better kilns to increase production. In 1864, the continuous-operation annular kiln, based on kilns used for making bricks, was introduced. Annular kilns were horizontal cylinders into which the material to be treated was fed continuously to the kiln. As the material flowed through the kiln, it was burned and then discharged. To improve the burning of the material in the kiln and improve the flow rate, the annular kiln was made to rotate at a few revolutions per minute. Lifting the material at a very slow speed dropped it through the flame from the burner. Rotating the kiln gave it the name *rotary kiln*, which was patented in 1877 and is still the major tool for producing portland cement today. The bottle kiln was adapted in the 1880s to continuous operation as a shaft kiln.

In 1898, F.H. Lewis wrote that the invention of the rotary cement kiln could be compared with the invention of the Bessemer process for steel production in terms of higher output and decreasing cost (Lewis 1898). He compared the production rates of the kilns in barrels per day:

- Intermittent bottle kiln (Aspdin's type): 15–30
- Continuous shaft kiln: 40–80
- Early rotary kiln: 120–180

But better kilns were not enough. The clinker still had to be ground to −100 μm in efficient fine-grinding mills. Making portland cement involved two stages of fine grinding. In the first stage, cement raw material containing 85%–90% limestone and 10%–15% clay and minor minerals was ground either wet or dry to a fineness of about 80% passing 150–200 mesh to make kiln feed. In the second stage, the kiln product—cement clinker—was ground dry to the fineness that, when mixed with water, aggregates, and gypsum and allowed to set, formed a hard cement.

Portland cement is ground to a fineness necessary to obtain its specified strength in a specific length of time. The fineness is defined by the Blaine surface area. For type 1 cement, a Blaine surface area of 3,060 cm^2/g is required to obtain full strength in 28 days, and for type 3 high-early-strength cement, a Blaine surface area of 4,950 cm^2/g is required to obtain full strength in 7 days.

In some parts of the world cements of various colors are made. The coloring agent can either be added to the finishing mill feed to obtain accurate blending in the grinding mill or can be added to the concrete mixer with the sand and aggregates. To prevent discoloration from the fine iron from abraded grinding balls and mill liners, either natural or manufactured pebbles are used as grinding media instead of iron or steel balls.

Cement clinker is always ground dry and must be kept dry until mixed with water to form concrete. The grindability (work index) of cement clinker is relatively constant even from different plants. Cement clinker stored over the winter in cold climates requires less energy for grinding than fresh clinker, particularly when the sides of the storage area are open. The change in power required is probably due to the clinker cracking during the cold weather with continued thawing and freezing. Stored clinker when ground, however, makes poorer quality cement. In some cement plants located in cold climates, clinker is made and stored at the start of winter when there is a much slower demand for cement, and the kilns are then shut down for the winter. In the spring clinker is available for grinding until the kilns are operating after maintenance and start-up time.

Cement kiln feed can be ground wet or dry. With the increase in demand for portland cement, the emphasis has been to reduce operating costs and the trend has been to dry grinding. Wet kiln feed contains 20%–30% by weight free water that has to be driven off in the kiln, which increases the fuel consumption in the kiln. Wet grinding is usually done in two-compartment, open-circuit mills to eliminate the water that is added to the classifiers if closed circuits are used. The grinding media in the first compartment can either be large balls or grinding rods. The use of rods in the first compartment was developed during the 1960s to give a more suitable feed for the open-circuit, fine-grinding, ball-milling compartment. It spread to other applications of multiple stages of wet grinding, such as grinding bauxite in a caustic solution to be fed to an alumina process plant.

Cement kiln feed is much less abrasive than cement clinker, which is the kiln product. With the use of wear-resistant materials in vertical-roller-mill grinding chambers and integral air classifiers from which the oversize is returned to the mill, vertical roller mills are becoming popular machines for dry-grinding cement kiln feed.

In the late 1950s and 1960s, with the introduction of single-stage ball mills for clinker grinding and the change to dry-grinding ball mills for cement kiln feed, the use of the same size ball mills for both purposes entered the picture. Often two ball mills of the same size as three clinker-grinding mills would be installed. The air separators would be the same size but the dry mills grinding cement raw material would be modified so that hot air would dry the ball mill feed before it entered the ball mill. Limestone is considerably less abrasive than cement clinker so ball life and liner wear in raw material grinding mills was very low. The use of ball mills for both clinker grinding and raw grinding decreased as the diameter of larger ball mills became too large for shipping to cement plants. Multiple-compartment mills capable of drawing more power than the same diameter, shorter length, single-compartment mills returned.

The need for more energy for grinding clinker and the limits on the size of mills that cement plants wanted to install led the way to the installation of high-pressure, double-roll crushers to crush cement clinker to ball-mill feed size. With better wearing parts and the internal classifiers, the same opportunity may be available to vertical roller mills.

The dry-grinding closed circuit used for grinding cement raw material caught the attention of the steel industry. A few steel mills started buying on the open market iron ore concentrates and high-grade iron ores that were too fine for direct feed to blast furnaces. They built pelletizing plants to make pellets from mixtures of concentrates and high-grade fines to feed to their blast furnaces. The experience of the cement industry in using dry-grinding closed circuits for making cement kiln feed was applied to dry-grinding iron ore concentrates and fine ore to pelletizing feed, which was about 80% passing 325 mesh.

The factors for converting wet-grinding energy—calculated from grindability tests to dry grinding, from closed-circuit to open-circuit grinding, and comparing grinding media and liner wear between wet and dry grinding—were developed from the grinding of cement raw material. Limestone, the main ingredient in portland cement, does not vary widely between plants and became the base for making these comparisons.

The cement industry's important contributions to fine-grinding technology were

- Establishing the tumbling mill as the most efficient fine-grinding machine available, an industry position that continued for more than 100 years
- Introducing the concept of closed grinding circuits in which dry mills were used in combination with air classifiers

The fineness of grind for the feed to cement kilns is about the same for all cement plants, as is the chemistry of the feed, with the base ingredient being limestone. The fineness of the grind for cement clinker is the surface area that meets the specifications for strength and days to set. Cement clinker is the product from the cement kiln and is, within narrow limits, uniform in grindability.

THE CYANIDE PROCESS

The discovery of fine-grained gold ores on the Witwatersrand in South Africa in 1886 highlighted the importance of finding a new method to extract the gold, because the losses with the amalgamation process used at the time were excessive. Chemists had known for 100 years that gold could be dissolved in potassium cyanide. A patent was taken out on the cyanide process in the United States in 1867, but neither the process that was the subject of that patent nor others taken out up to 1885 were used on a commercial scale. Recovering all the gold from the solution was a difficult problem to overcome.

In 1887, J.S. McArthur and R. and W. Forrest, working in a laboratory in Glasgow, Scotland, observed that a cyanide-leaching process followed by a zinc-precipitation process dissolved the gold from ores and ensured that there was a high recovery of gold from the pregnant solution (Wilson 1908). The cyanide process proved to be satisfactory on a small scale in Ravenswood, Queensland, Australia (Lougheed 1985), but its real test came when it was used to extract the gold from tailings on the Rand. It was very successful, as indicated by the value credited to the cyanide process by the South African gold industry, which increased from $6,000 in 1890 to $6 million in 1893 (Dorr 1936).

THE FLOTATION PROCESS

The background to the flotation process was similar to that for the cyanide process. The reserves of ores containing coarse-grained minerals were dwindling, and ores containing fine-grained minerals—which were difficult to extract by the processes of the time—needed to be mined. The cyanide process was unsuitable for ores containing lead, zinc, and copper, all of which occurred as sulfide minerals. Gravity-separation processes gave mediocre results with ores containing these minerals in the form of fine grains. By 1900, the problem of mineral losses in tailings had become serious.

The rich and large lead-silver-zinc ore deposit at Broken Hill in Australia focused attention on this problem. This deposit had the form of a giant vertical boomerang with its ends pointing down and its apex breaking the surface. The oxidized ores at the surface were coarse grained and the minerals could be extracted easily. But the deeper primary ores were a serious problem, because the lead, silver, and zinc sulfides in them were fine grained, and finer grinding was needed to liberate them for concentration. Mining began at Broken Hill in 1885, and by 1893 the high-grade oxidized ores were nearing exhaustion. The future of the town of 20,000 people depended entirely on the discovery of processes to extract sulfide minerals from the fine-grained primary ores.

After years of expensive effort, flotation proved to be the solution. It worked because sulfide minerals have different surface properties from oxides, and, if the pulp chemistry is right, air bubbles will attach to sulfides and float them to the surface while oxides are not affected. Tube mills, which had been used successfully in the gold industry, were adapted to grinding sulfide ores to liberate the fine minerals and to make small particles that air bubbles could lift.

The cyanide and flotation processes remove metallic minerals from the ore. These minerals have been liberated from the ore by the fineness of grind, which is determined by the size of the mineral particles in the ore. The types of rocks that contain minerals vary widely in grindability and abrasive characteristics. The size of the mineral particles varies with the type of minerals in the ore body. Testing is required for each ore type and deposit to determine its grindability and abrasive characteristics; no rules of thumb or comparisons are available, as there are for grinding cement raw material and clinker. In the manufacture of portland cement, size reduction is an integral part of the process, particularly in the grinding of the clinker. For the cyanide, flotation, and other mineral concentration processes, size reduction is part of the feed preparation stage, not part of the process. The need to properly prepare the feed led to the development of wet closed-circuit grinding in the tumbling mills discussed in this chapter.

THE EVOLUTION OF TUMBLING MILLS

Tumbling mills are built for either continuous or batch operation. In continuous mills, feed enters one end and broken product leaves the other. In batch mills, the material to be ground is charged to the mill, and the mill is closed and run until the material is ground. The mill is then stopped, the load in the mill is dumped out, and another load is put into the mill. Ball mills use steel, cast iron, or sometimes ceramic balls as the grinding media. Long ball mills called "tube mills" consisted of rotating drums partially filled with rocks or pebbles that were lifted to the top of the drum, then tumbled freely to the bottom. The tumbling grinding media would break the material being fed to the drum to the desired fineness. The fineness of the material discharged from the drum could be made coarser or finer by increasing or decreasing the feed rate to the drum. Before the development of the cyanidation process for gold extraction and the flotation process for sulfide mineral extraction, a few installations of small tumbling mills ground metallic mineral ores.

Tube mills were used to grind gold ore in mines on the Rand in South Africa in the early 1900s. The first mills used local pebbles for grinding media, but grinding pebbles imported from Belgium and Normandy were found to be harder and have a longer service life than the local ones. When the cost of these imported pebbles became prohibitively high, the South Africans began testing pebbles removed from their ores. After World War II, they began to test autogenous grinding, in a process that became known as primary autogenous grinding and led to secondary autogenous grinding, which uses cast iron or forged steel grinding media instead of pebbles extracted from the ore. The heavier weight of the iron and steel ball increased the power drawn by the mills, which in turn increased the mill capacity. By the 1930s, many of the South African gold mines were using ball mills, which led to the use of large-diameter, long tube mills with open discharge ends at mill speeds to 90%–95% of critical speed. The discharge ends of the mills were carried on shoe bearings fitting around the mill shell. These types of mills are still used in the gold mining areas of South Africa.

The early tube mills used to grind mineral ores had diameters of 0.9–1.5 m with the length about twice the diameter. By the end of World War I, the diameter of ball mills had increased to 2.4 m, and by the end of World War II the diameter of ball mills was 3.2 m. The size of the drive motors had increased from 10 to 600 kW, and the two- and three-compartment ball mills used in the cement industry required 933-kW motors. As the 21st century began, mill diameters had increased to 6.7 m with motor sizes up to 15 MW.

At the same time that pebble milling started in South Africa, gold and silver mines in Mexico, Honduras, Australia, and the United States started using pebble mills to grind their ores. These mines began using pebbles from their ores and relatively quickly converted to cast iron and forged steel balls. The native copper ore mines in the Lake Superior mining district of the United States began regrinding the tailings that still contained copper with pebble mills, which were eventually converted to ball mills.

Just before World War I, tumbling mills using rods 150 mm shorter than the length of the mill shell as grinding media were introduced. These rod mills could be fed ore that was approximately 12 mm larger than the feed to ball mills and could grind ore in open circuit to about 10 mesh, which made good ball-mill feed. Rod mills replaced double-roll crushers and covered the finer end of the range of product sizes available from cone crushers. Rod-mill ball-mill circuits improved the power efficiency of grinding circuits. After World War II, rod-mill ball-mill circuits were installed in new plants processing ferrous and nonferrous ores, but, beginning in the 1970s, two factors slowed the installation of rod mills. First was the availability of quality rods that would have an economic life as grinding media and would not bend or break prematurely. The grinding rods in a rod mill must stay straight and be parallel with the length of the mill. If rods bend or break prematurely, they disrupt the action of the rod charge and cause the rods to bend and disrupt the parallel rotation of the other rod. The rod charge becomes a tangled mess of bent and broken rods. The length of the rods should be at least 1.4 times the diameter of the rod mill to prevent tangling. The limit on the availability of quality rods was 6.7 m, which limited the diameter of rod mills to 4.57 m. Better quality rods became available late in the 20th century. The second factor was the use of primary autogenous and semi-autogenous rod mills to replace cone crushers and rod mills in size-reduction circuits.

TUMBLING MILLS IN THE 19TH CENTURY

Although ball mills were tested in the 1860s (Del Mar 1917), the main progress in fine grinding during 1870–1900 was in Germany. The German mineral industry had started to decline after hundreds of years of world leadership, but the cement industry was growing and with it the demand for fine grinding of clinker. The growth of chemical engineering in Germany during that time also created a demand for very fine particles for use in chemicals, paints, pharmaceuticals, and other products. Invention was the order of the day, so it is not surprising that

> *...in 1876 a machine was proposed by Gebruder Sachsenberg in which the screening device fixed on wooden frames surrounded the grinding drum and rotated with it. The material passing through the screen formed the finished product and was caught in a casing surrounding the machine, while the material retained on the screen was returned to the inside of the mill through special openings. (Scheibe 1993)*

The mill was designed by the Sachsenberg brothers in Roblau-on-Elbe and W. Bruckner in Ohrdruf and in 1885, a mill was built in H. Gruson's workshop in Magdeburg, Germany (Naske 1911). It was the first tumbling mill to gain reasonable acceptance in the industry. The Krupp Company, which was dominant in Germany at that time in manufacturing heavy steel equipment, acquired Gruson's company and built the Krupp-Grusonwerk mill. Figure 7.1 shows the Sachsenberg mill.

The mill consisted of a drum lined with steel plates with the leading edge of one liner sitting on the trailing edge of the next. The grinding media were steel balls 63–100 mm in diameter, and the steps in the plates kept them in motion and grinding clinker when the mill was rotating. Fine particles were discharged from the mill through apertures in the plates, and coarse particles in the discharge were screened and reentered the mill through slits between the plates.

The early Krupp-Grusonwerk mills were dry mills. Because cement production in 1885 was still small, progress with tumbling mills was slow. But rapid growth of the cement industry brought them to the fore, and by the end of the 19th century tumbling mills were being used to grind metalliferous ores to prepare them for the cyanide process. A few years later ball mills were being used to grind ores for flotation. Figure 7.2 is a photograph of three Krupp-Grusonwerk mills that were used for dry grinding gold ore in Kalgoorlie, Australia, in 1905. A problem with the Krupp-Grusonwerk mill was that the apertures in the drums through which the ground particles were discharged had to be large enough to minimize blocking, and this gave a rather coarse product, which

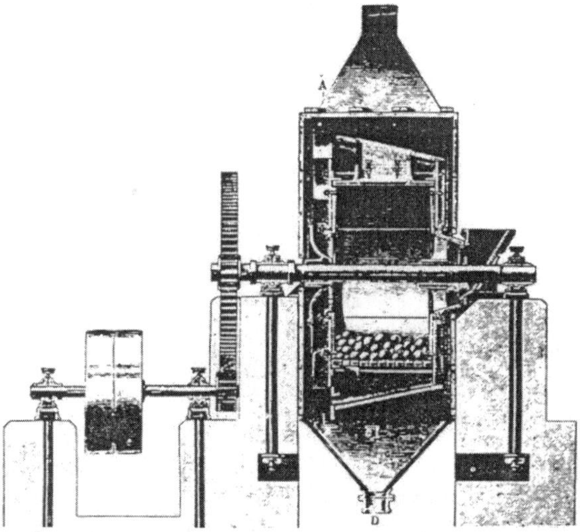

FIGURE 7.1 Sachsenberg mill, the first model of the Krupp-Grusonwerk mill (Tiggesbaumker 2001; reprinted by permission from Krupp Polysius, a Company of ThyssenKrupp Technologies)

FIGURE 7.2 Three Krupp-Grusonwerk ball mills grinding gold ore dry in Kalgoorlie, Australia, in 1905 (courtesy of John Tusa)

made substandard cement. In a later form of the mill, the product was discharged through an end trunnion, which improved the performance considerably. By 1900, a mill 1.2 m in diameter and 6 m long and using flint pebbles ground 3 tph of shaft kiln clinker to cement (Scheibe 1993).

The solution to the problem of large particles in the product was to operate the mill in closed circuit with an air classifier, collecting the fine particles from the classifier in one product and recycling back to the mill the large particles that were in the other product. The Askham air separator was patented by Mumford and Moodie in 1885. Figure 7.3 shows how an Askham separator was used in closed circuit with a Krupp-Grusonwerk

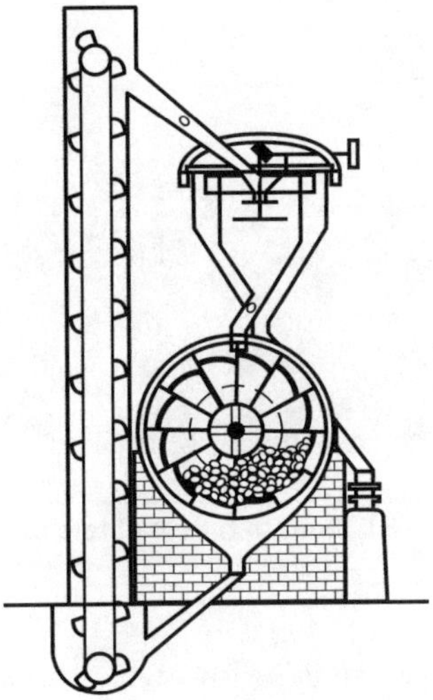

FIGURE 7.3 Krupp-Grusonwerk mill in closed circuit with an Askham air separator (West 1910)

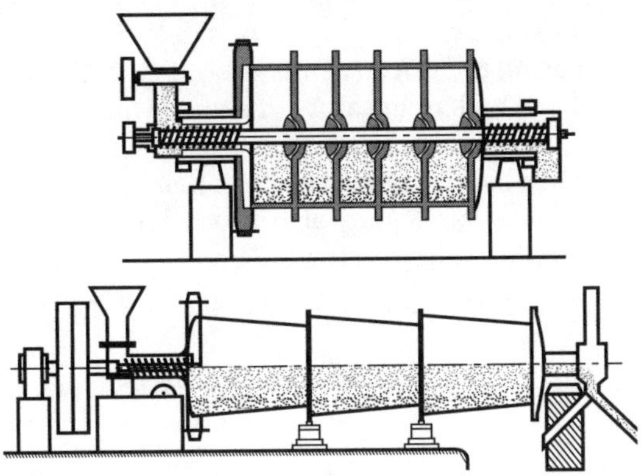

FIGURE 7.4 Ideas for continuous ball mills in the late 19th century (Scheibe 1993; reprinted by permission of Klaus Graichen)

mill. This type of circuit started a trend that eventually changed industrial grinding processes from mainly open circuit to mainly closed circuit. But it would not become common practice for many years. By 1900, tumbling mills were competing successfully for grinding cement clinker with Griffin-type centrifugal pendulum mills.

The inventors of the day had many ideas about ball mills, and Figure 7.4 shows some methods that were considered for continuous ball mills. Ultimately, the simplest idea proved to be the best: grind in a cylinder with ore entering one end and leaving the other.

TUMBLING MILLS FOR MAKING CEMENT

By 1890, 100 Krupp-Grusonwerk tumbling mills were operating in cement plants, and other companies were manufacturing them. One company that contributed much to the technology of tumbling mills was F.L. Smidth and Co. of Denmark. This company had been established as a one-man consulting engineering business by Frederik Laessoe

FIGURE 7.5 Transporting an early F.L. Smidth tube mill to a cement plant in North Africa (Smidth 2000; reprinted by permission)

Smidth in 1882 and had grown rapidly through contracts to build cement plants. In 1893, the firm acquired the rights to a tube mill from a French inventor and sold it worldwide after redesigning it. The problem was that mills were so large and heavy that it was often difficult to transport them to plant sites. One solution is shown in Figure 7.5.

One approach to grinding clinker involved using a short tumbling mill, called a ball mill, to break the fresh clinker to the size of grit or sand and a long tumbling mill, which was called a tube mill, to grind the grit to cement. Both mills used trunnions to discharge the ground products. Short ball mills and tube mills made by Krupp are shown in Figure 7.6.

The success of tube mills led to finished cement being produced in shorter tube mills operating in closed circuit with air separators or in longer tube mills operating in open circuit. The details of one mill-separator circuit were

- The mill was 1.6 m in diameter and 3.6 m long, and the separator was 2.8 m in diameter.
- The rotary kiln clinker was fed to the circuit in pieces up to the size of hazelnuts at a rate of 4.5 tph.
- The circulating load was 150%–200% and the finished cement contained 16% +90 μm particles.
- The power consumed by the mill, separator, and elevator was 82 kW (110 hp; West 1910)

Next the short mill and the tube mill were combined into one machine, shown in Figure 7.7, and mills of this type were made in Germany. It can be regarded as the original two-compartment mill.

> *The material to be ground is fed into the hollow trunnion in the ordinary way, and in the first part of the mill is acted upon by a number of steel balls of varying sizes, by which it is reduced to a grit by the time it reaches the division between the two portions of the mill. Here it falls on to a steel plate provided with suitable apertures, which only allows the sufficiently reduced material to pass, the residue being returned to the ball mill part for further reduction. The material which passes the openings in the steel plate is carried forward into the ball mill portion for further reduction. (Butler 1913)*

The Gates Iron Works of Chicago was a leading manufacturer of machinery for the mining and cement industries. Figure 7.8 is a photograph from a cement plant of the early

FIGURE 7.6 (a) Short tumbling mill (ball mill) used for the initial grinding of clinker and (b) longer tumbling mill (tube mill) used for final grinding to finished cement (Butler 1913)

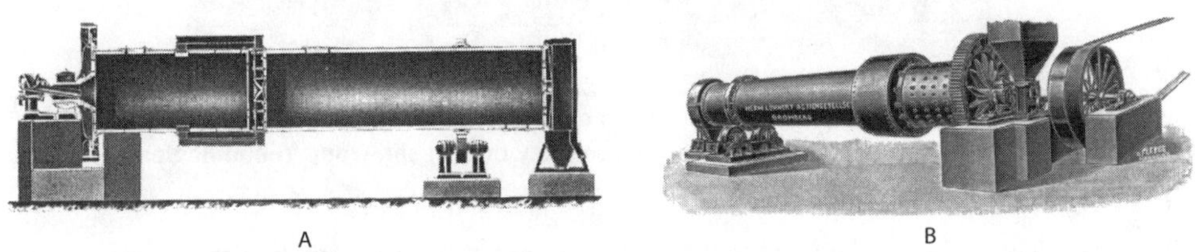

FIGURE 7.7 Ball mill and tube mill in one machine (West 1910)

FIGURE 7.8 Gates peripheral discharge mills and tube mills in a cement plant in the early 1900s (Peterson 1976; reprinted by permission from Milwaukee County Historical Society)

1900s showing Gates peripheral discharge mills in the foreground and tube mills for fine grinding in the background. The peripheral discharge mills were similar to the mills with peripheral screens developed by Krupp in Germany. By the early 1900s, Krupp mills were trunnion discharge mills. A mill in a cement plant in 1906 is pictured in Figure 7.9.

TUMBLING MILLS FOR GRINDING ORE

After 1900 the grinding of Portland cement raw material and of cement clinker required large numbers of tumbling mills. This was also the heyday of gold mining. The old stamp mills that were used in great numbers for grinding gold ore did not grind sufficiently fine to liberate all of the gold, and the new tube mills were

FIGURE 7.9 Trunnion discharge Krupp ball mill in a cement plant in 1906 (Tiggesbaumker 2001; reprinted by permission from Krupp Polysius, a Company of ThyssenKrupp Technologies)

> *installed following the stamps. After 1910 larger diameter tumbling mills with larger grinding media were developed. These could receive the finer crushed ore directly and the inefficient stamp batteries were gradually eliminated. ...The first tumbling mills, or tube mills went into operation (on the Rand) in 1904. (SME–AIME 1985)*

By 1900, ball mills had been tested on the gold ores at Kalgoorlie, Australia, but the results were not always satisfactory.

> *Ball mills had for a long time a bad reputation...due to the fact that the excellent mills of this type built by Krupp had been imitated by local makers with unsatisfactory results; since this has been recognised they are again coming into use. (Schmeisser 1898)*

A paper published by the Institution of Mining and Metallurgy in London gave a glimpse of how wet-grinding circuits worked on gold ores in 1904 (Rose 1906). The author reported that a circuit in the Hannan's Star mill in Kalgoorlie consisted of a tube mill 1.2 m in diameter and 5 m long operating in closed circuit with a hydraulic classifier. The feed to the circuit was 38 tph of sands from a classifier, and the flow rate of the secondary hydraulic classifier sands back to the tube mill was 230 tph. The circuit feed and product sizings are given in Table 7.1. These data were representative of operating conditions at the time and show there was (1) efficient desliming by the hydraulic classifier, (2) a very fine classifier overflow stream that generated a large circulating load, and (3) overgrinding of the circuit product.

There was a clear need for better classification in wet-grinding circuits, a problem that was to be solved within the next few years.

The most spectacular changes that electricity brought to tumbling mills were the increases in mill and motor sizes (see Table 7.2) and in mill capacities. In 1900, few motors for tumbling mills were larger than 5 kW, but by 2000 electric motors were as large as 20,000 kW. In 1900, a few tons of ore could be ground per hour, but in 2000 a common production rate of a circuit containing a large mill was many hundred tons per hour.

Scale up of tumbling mills involves a number of factors as demonstrated by the scale. The power drawn by ball mills operating at the same (1) percent of volumetric

TABLE 7.1 Sizing distributions of the feed to and product from a tube mill–hydraulic classifier secondary grinding circuit in the Hannan's Star mill, Kalgoorlie, in 1904

Mesh	+40	40/60	60/100	100/150	–150
Feed	20.5	43.8	28.1	7.6	—
Product	—	—	0.7	4.0	95.3

Source: Rose 1906.

TABLE 7.2 Increases in dimensions and motor size of ball and autogenous/SAG tumbling mills in the 20th century

Ball Mills				Autogenous/SAG Mills			
Year	Diameter, m	Length, m	kW	Year	Diameter, m	Length, m	kW
1909	1.2	2.1	11	1948	3.05	1.01	75
1912	1.9	2.3	41	1960	6.7	2.1	1,119
1927	2.4	2.4	168	1969	9.8	4.27	5,220
1940	3.05	2.8	447	1979	10.4	5.2	9,321
1963	3.9	5.5	1,491	1987	11	5.2	11,186
1970	5.6	6.4	3,169	1994	11	5.8	13,423
1990	6.1	9.3	5,593	1996	12	6.8	19,388
1997	7.3	10.5	10,440	1997	12.2	6.1	20,000

Source: Taggart 1927; SME–AIME 1985; authors' personal notes.

loading, (2) percent of critical speed, and (3) length of grinding compartment varies as the ratio of the mill diameter is raised to the 2.3 power $(D1/D2\ )^{2.3}$. This applies when the power drawn by one mill diameter is known. This keeps the three key variables in the mill power equation at the same level with only mill diameter and power as variables. The mill volume varies as the diameter squared. As the mill diameter increases, the volume of the mill available per unit volume of the feed decreases. As ball mill diameter changes for a constant percent of critical speed, mill speed in revolutions per minute changes at inverse ratio of the square root of the mill diameter. The volume of the media in the mill is a function of the ratio of mill diameters squared. As the mill diameter increases the mill speed in revolutions per minute and the number of balls per unit of feed decrease, so the number of ore and media contacts decreases. As ball mills become larger in diameter, the largest mills become more inefficient—shown by either a decrease in feed rate to produce the specified grind or by a coarser grind at the designed feed rate. These scale-up factors result in the same manner in all tumbling mills. Mill power that is mill capacity is directly proportional to mill length.

Tumbling mills brought seminal change to fine grinding because they could grind pebbles to fine particles quickly and with much smaller use of energy than was previously possible. But they were not the panacea for all fine-grinding problems; for example, they are inefficient for breaking particles smaller than about 20–30 μm, because very small particles, like liquids, flow away from colliding balls. Changing the movement of the balls from tumbling to stirring reduced the problem of very fine grinding in ball mills. Stirred ball mills are discussed in Chapter 8.

THE RISE OF BALL MILLS

Ball mills became the dominant fine-grinding machine in the 20th century. By 1900, the ball mill patent litigation in France had ended, and inventions to improve fine-grinding processes were in full flood. Inventing a machine is one thing, selling it another. One event that went far to establish the credibility of wet ball mills as fine-grinding machines was the installation, in 1910, of 64 Hardinge conical pebble mills in the retreatment plant of the Calumet and Hecla copper mine in Michigan's Upper Peninsula.

Harlowe Hardinge was a consulting mining engineer who observed that large and small rocks moved at different rates when moving down an incline. It occurred to him that if a mill had a cylindrical section at the feed end attached to a long conical section at the discharge end, the rocks that were the grinding media would arrange themselves in the mill according to size, with the coarse media in the cylindrical section and the fine media spread along the conical section (*Crushing and Grinding* 1931).

This would improve the efficiency of grinding by breaking the coarse feed particles early in the mill, with coarse media and fine particles later in the mill with fine media. The same concept is used with classifying liners today. Figure 7.10 shows the mill that Hardinge built.

The Hardinge mill had trials in Pennsylvania and Missouri with pebbles as the grinding media, but it was at the renowned Calumet and Hecla Mine where it established a reputation as the most suitable machine for grinding hard metalliferous ores. Native copper had been mined in this Lake Superior region since prehistoric times (Drier and Du Temple 1961). The deposits were large, and several companies were mining ores containing native copper and primary copper sulphides in the 19th and 20th centuries. Milling practice at the Lake Superior mines in the 1850s was to break large boulders containing the copper by hand and then to use Blake jaw crushers and stamp mills for further breakage and to concentrate the copper minerals on jigging tables. This practice continued for decades, and the recovery of copper was always poor.

In 1908, the management of the Calumet and Hecla Mine, which had become the main operating company in the region, installed a processing plant to extract copper from 37 million tons of tailings that had accumulated over 50 years. These tailings contained 5 kg (11 lb; 0.5%) of copper per ton of ore (Benedict 1955), but they were hard and tough and had to be ground from 6 mm to −0.2 mm. It was the practice at the time when building new plants to run a competitive test between machines from rival manufacturers. A 12-month trial was carried out with three different mills—a tube mill, a Chilean mill, and the Hardinge conical mill. The demanding test was won by the Hardinge mill, and the result was that 64 open-circuit mills were installed in a new tailings treatment plant, each 2.6 m in diameter with a 0.5-m cylindrical section and using flint pebbles as grinding media. They are shown in Figure 7.11.

The use of the Hardinge mill increased copper recovery immediately to 75%, and, when steel balls replaced the flint pebbles a few years later and increased the energy that could be used in grinding, this recovery rose to 90%. By then, flotation was also being used, so how much of the total increase was the result of flotation and how much resulted from finer grinding is not known. It is sufficient to say that both the tumbling mill and the type of media used contributed to a substantial increase in the recovery of copper.

The Role of the Mill Manufacturers

For most of the 20th century, equipment manufacturers brought the main improvements to the technology of fine grinding. The Allis-Chalmers Company, for example, with its research programs and extensive publications, was a leading manufacturer of crushing and grinding machinery and one of the main contributors to better grinding technology. The Allis-Chalmers Company, then, is a good focus for a discussion of the problems confronting the mill manufacturers and how they were solved.

The company that became Allis-Chalmers was started in Milwaukee, Wisconsin, in 1847 as the Decker and Seville Company, which made millstones for grain mills from flint and chert exported to the United States from the coasts of Denmark and Normandy.

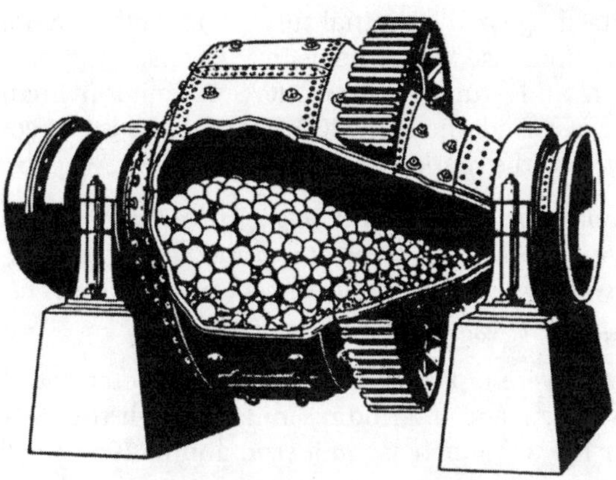

FIGURE 7.10 Hardinge conical mill, an early wet-grinding mill designed to improve efficiency by using segregation of the balls. The cylinder of this mill had a peripheral speed of 195 m/min, and the larger 125-mm balls remained in this section. The peripheral speeds of the conical section at the discharge end ranged from 120 to 75 m/min and the balls graded from 62 to 87 mm (Seymour 1924).

FIGURE 7.11 Sixty-four Hardinge conical mills in the new Calumet and Hecla tailings retreatment plant in 1910 (Benedict 1952; reprinted by permission from Michigan Tech University)

It was purchased by E.P. Allis & Company, which then grew because of strong management and innovative ideas. In 1901, E.P. Allis combined with other companies to form the Allis-Chalmers Company. Others in the new company were the Gates Iron Works, a large supplier of mining equipment, including tumbling mills, and the Fraser & Chalmers Company, which first made grinding mills in 1896. The merger gave Allis-Chalmers a complete line of comminution machinery that it supported by making steam engines, electrical motors, motor-starting equipment, and belt and pulley drives.

Geographically, Allis-Chalmers was well-placed to be a leading supplier of equipment for the resource industries. There were copper and iron deposits in the Upper Peninsula of Michigan where Hardinge had his initial success, as well as in northern Wisconsin and Minnesota and in the immense wheat fields in Wisconsin and other midwestern states. With companies such as Hardinge, Marcy, KVS, Dominion Engineering, and Denver Equipment becoming involved, the early 20th century saw many competitors for the U.S. grinding mill market, and the market for tumbling mills grew quickly.

> *In the ore concentrators of the United States (in 1938) there are about 650 ball, rod or pebble mills operating or ready to operate and these have about 100,000 connected horsepower. For grinding in cement plants there is about three times this connected power (Coghill and DeVaney 1938).*

Tube mills in 1905 were typically 1.3–1.9 m in diameter and they were powered by motors up to 22 kW. The trend in grinding mills through the 1920s, the Great Depression, and World War II was to increase mill size, and in 1951 the largest available mills listed in the Allis-Chalmers bulletin were

- Wet-grinding rod mills: 3.2 m in diameter, 3.7 m long, 522-kW motor
- Wet- and dry-grinding ball mills: 3.2 m in diameter, 4.9 m long, 746-kW motor
- Dry-grinding mills with more than one compartment: 2.9 m in diameter, 10.3 m long, 932-kW motor

Although large mills may have received most of the publicity, the use of smaller mills in remote areas often required much ingenuity. Figure 7.12a shows a small mill in a mountainous area of the Philippines that was driven by a vertical water wheel and ground a few tons of ore per day. This type of small operation has always contributed to the economy of many countries. Figure 7.12b is a photo of large grinding mills in a molybdenum concentrator in the United States. These mills ground thousands of tons of ore each day and were equipped with many monitoring and control devices.

Early in the ball mill era, choosing a mill and motor for a specified feed rate, product size, and ore grindability was found to be problematic. As a result, Allis-Chalmers established a research laboratory to work on grinding, which was led by Fred Bond. Bond's work was an important factor in making Allis-Chalmers the leader in grinding technology for much of the 20th century. His name will always be associated with ball milling.

As the diameter of tumbling mills increased, the manufacturers required advanced engineering approaches to solve design and manufacturing problems. We discuss these approaches later in this chapter.

Polysius or Krupp-Polysius mills have been mentioned several times in this history. As background, Andreas Polysius set up a mechanical workshop in Dessau, Germany, in 1859, and in 1870 he started making mills for grinding construction materials. By 1898, the company was making rotary kilns for the manufacture of cement and some years later was building complete plants. It continued to expand and in 1971 was purchased by Krupp.

ROD MILLS

With the development of ball mills, studies were made to find improved means to produce finer feed for ball mills. Double-roll crushers, gyratory crushers, and cone crushers were developed for this role. Using steel rods as the grinding media in tumbling mills was found to be an efficient means to make ball mill feed and to make a product finer than could be made in a crusher but coarser than could be made in a ball mill. The principal use for rod mills, which were invented in Germany around 1870, was to make ball mill feed in a wet-grinding rod mill–ball mill circuit.

Rod mills came into use in the United States beginning in 1910 and eventually became recognized as efficient fine crushers. "The installation of a rod mill at Sudbury,

A

B

FIGURE 7.12 (a) Small ball mill in the Philippines grinding gold ore and driven by a breast-shot water wheel. The mill is partly hidden by the water wheel. (courtesy of Bruce Ashton, consultant, Adelaide) and (b) large tumbling mills grinding molybdenum ore in the United States (courtesy of Metso Minerals)

FIGURE 7.13 Tangled rod load in a 3.8-m by 5.2-m rod mill grinding copper ore

Ontario, by Mining and Smelter Supply Company of Denver, Colorado, was the beginning of the end for stamps in mines processing ores containing precious metals" (Bond 1964).

In 1916, Allis-Chalmers built a rod mill (2 m in diameter by 4 m long) for the Saint Joseph Lead Company in Missouri, to be installed at the Bonne Terre mine where a coarser feed was required for shaking tables. Crushing rolls and rod mills were eventually used in all the company mines to prepare feed to shaking tables. In the 1950s, rod mills 2.9 m in diameter by 3.9 m long replaced the roll crushers.

With the increase in the diameter of rod mills, problems occurred in the shorter rod mills with rods tangling. Figure 7.13 shows an example of a tangled rod charge. When the rods were short relative to the diameter, some were lifted toward a vertical position, and continued tumbling caused tangling. This led to the arbitrary rule that the length of rods should be at least 1.4 times the mill diameter inside new shell liners, although this was not enough at times. Because the maximum length of high-quality rods was 7.1 m, this set the limiting diameter for rod mills at no more than 5.2 m. High-quality rods can now be manufactured to a length of 8 m, and mills can be built that are 6 m in diameter and 8 m long, with a 4,500-kW motor.

Improvements in the performance of fine crushers and the introduction of primary SAG mills have reduced the use of rod mills in crushing and grinding circuits.

Ball Milling after World War II

The demand for iron, copper, and other strategic metals during World War II was enormous, and their resulting production depleted the available reserves of high-grade ores. After the war, large volumes of metals and cement were needed to rebuild industry and infrastructure, and a large increase in grinding capacity was required. In the United States, an increasing amount of iron had to come from taconites, which are very hard, abrasive siliceous magnetite ores that contain 20%–30% iron. Magnetite extraction involved rod and ball milling with interstage magnetic separation, and the concentrates were 80% passing 38 μm. The early plants used mills that were 3.2 and 3.8 m in diameter, and even then there was a hint of a problem with rod tangling, which became significant in later plants that used mills up to 4.9 m in diameter. It became apparent that 4.5–4.9 m was the limiting range for rod mill diameters because of rod tangling. No limitations for ball mill diameters were encountered in iron ore concentrators.

Expansion of copper ore mining began in Arizona and Chile in the mid-1950s and accelerated in the 1960s worldwide. Single-stage ball mill and two-stage rod mill–ball mill circuits were both used. Rod mill growth stopped at 4.9 m, but ball mill growth continued up to 7.7 m in single-stage circuits and as the second stage in SAG mill–ball mill circuits. Efficiency problems, which started at about 5.8 m, have never been explained satisfactorily. However, because the efficiency issues vary in magnitude, they are probably caused by a combination of scale-up factors, ore size and grindability, and the total mill throughput. In both the mining and cement industries in the 1960s and 1970s, increased demand was seen for larger grinding mills that drew more power and had higher grinding capacities. Rod mill diameters increased to 4.9 m, ball mill diameters rose to 6.5 m, and multiple-compartment mills increased to 4.5 m in diameter and 13 m long. There were even more spectacular increases in the sizes of large-diameter, short-length autogenous and SAG mills (Table 7.2).

In the 1950s, the use of hydrocyclones for classification changed the design of grinding circuits almost overnight, allowing for the use of larger diameter mills. The rise of hydrocyclones is discussed in Chapter 9. Briefly, hydrocyclones displaced rake and spiral classifiers in the 1960s, which was timely because mechanical classifiers had just about reached their size limit. Further use of them with larger mills would have required the installation of parallel units requiring expensive space and involving complex material-handling problems. Hydrocyclones solved these problems and reduced the capital cost.

Multiple-Compartment Tube Mills

Starting in the late 19th century, particularly for grinding in the cement industry, tumbling mills were built in which a short ball mill was attached to a tube mill to make a single mill with a division head separating it into two compartments. Large balls were charged to the coarse grinding compartment. Worn balls in the first compartment would pass through the division head and flow into the second compartment where they were needed for fine grinding. To maintain the volume of the grinding media in the fine-grinding compartment, it was necessary to periodically shut down the mill to add small balls to the second compartment. Balls were added to the first compartment with the mill feed. The design and manufacture of multiple-compartment ball/pebble mills became the heart of manufacturing tumbling mills to grind cement clinker and cement raw material.

To maintain the flow of material through very long mills, it was necessary to add a second division head to lift the material, increasing the flow gradient through the mill. This made the mill a three-compartment mill. In the 1920s and 1930s, many combinations were used for dry, multiple-compartment mills grinding cement clinker. To increase the amount of impact grinding, the diameter of the first compartment would sometimes be made larger than the diameter of the second compartment. This caused stress cracking in the tapered section of the shell joining the shells of the first and second compartments. With closed circuiting there were many combinations of open- and closed-circuit arrangements. Ground clinker was peripherally discharged from the first compartment and then fed to either screens or air classifiers, with the oversize returned to the first compartment. One example of this was to peripherally discharge the closed-circuit first compartment with the discharge going to a screen. The oversize went back to the feed end of the mill. The fines were fed to the second compartment either by a scoop feeder through openings in the shell into the division head between the two compartments of the mill or through a spout feeder at the discharge end trunnion. If the oversize from the screen was fed to the second compartment through the discharge diaphragm, there would be a separate peripheral discharge on the second compartment side of the division head. The second compartment could either be open- or close-circuited with an air separator. The design confusion was solved as multiple-compartment mills grinding cement clinker were standardized as closed-circuited mills with all of the mill discharge being fed to classification systems, and the oversize from the classifier being returned to the feed end of the mill. Closed circuiting of the first compartment was discontinued.

In the 1950s, the use of single-compartment, large-diameter ball mills in closed circuit with an air separator was developed. Still used today, this circuit is the same, in principle, as the single-stage ball-mill circuit used in mineral concentrators. The trend in large tumbling mills for grinding cement clinker is toward installing two-compartment tube mills in closed circuit with an air separator. The increase in energy use comes from the increase in mill length rather than an increase in mill diameter.

Work continues on improving the metallurgy of the wearing parts of the vertical roller mill so it can be used as a primary mill for grinding cement clinker. In the 1980s, for example, horizontal double-roll crushers with hydraulic high-pressure roll retainers proved to be successful in crushing cement clinker to make tube mill feed.

With the Krupp-Grusonwerk mill that was first used for grinding clinker to finished cement, the mill contents were fine and the product that passed through the circumferential screen was fine enough to give a consistently good cement when the clinker was soft. But when the clinker was hard, the product was coarser and the cement made from it was often unsatisfactory. The growing demand for good-quality cement highlighted the need for better equipment, and lawsuits in the 1890s emphasized the fact that ball mills were becoming the preferred machines for making very fine particles. Clinker grinding became a two-stage process carried out in a single-tube mill with two compartments (see Figure 7.15) separated by a partition that had four functions:

- Retained large balls in the first compartment to break large lumps of clinker
- Kept small balls in the second compartment to produce very fine particles
- Allowed the air, which both dried and classified the limestone or clinker, to flow through the mill freely
- Lifted the discharge from the first compartment to the horizontal center of the mill to be fed to the second compartment to give a gradient to the flow of material through the mill

TUMBLING MILLS, PATENTS, AND LAWSUITS

In the mid-1890s, lawyers became interested in tumbling mills. In 1891, a patent specification was lodged with the Imperial Patent Office of the German Reich (Scheibe 1993) in which priority was claimed for the use of solid metal sections rather than screens on the cylindrical part of the mill. The purpose of the invention was to eliminate the problems in operation caused by the screens. Figure 7.14 shows the cylindrical and conical mills described in the patent. Patents for tumbling mills were ignored until 1897 when very fine grinding started to become important. The Krupp Company and the Luther Company had been making tumbling mills in Germany for some years by then, so they strongly defended their case when the claim was made that they infringed on the 1891 patent. A 1-year lawsuit took place during which decisions about the validity of the patent were made, which were reversed on appeal, but eventually the lawsuit ran out of steam as the patent's expiration date approached.

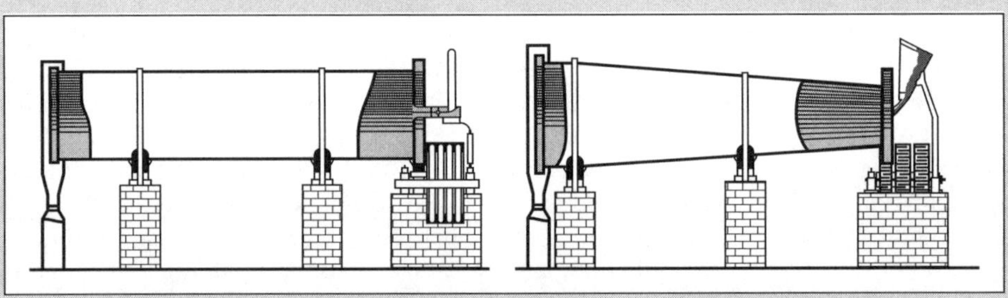

FIGURE 7.14 Tumbling mills described in the 1891 German patent (Scheibe 1993; reprinted by permission of Klaus Graichen)

FIGURE 7.15 Allis–Chalmers two-compartment cement mill of the 1920s (Peterson 1976; reprinted by permission from Milwaukee County Historical Society)

Tube mills with two and occasionally three compartments became the workhorses for grinding cement clinker and other dry materials during the 20th century and will be used extensively for dry grinding for many years. Tube mills have increased greatly in size to meet the rising demand for cement; mills are now as large as 5.5 m in diameter by 16.5 m long, and powered by 7,000-kW (10,000-hp) motors (Duda 1987).

Figure 7.16 shows a modern, high-capacity two-compartment mill. Mill capacities have increased tenfold since 1950, from 35 to 350 tph. The engineering techniques for design and construction that are now available mean that mill diameter and length can

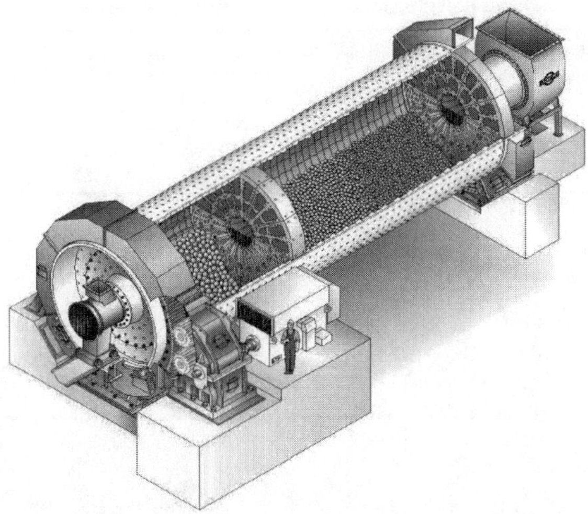

FIGURE 7.16 Modern two-compartment tube mill (reprinted by permission of Krupp Polysius)

be increased to meet any required capacity. Two-compartment tube mills modified for intermediate classification between the compartments are finding increasing use for grinding refractory gold ores that must be roasted to enhance the efficiency of leaching. These types of tube mills are described later in this chapter. As two-compartment mills have become larger and their capacities have increased, attention has turned to the internal fittings of the mills such as the ball loading, the division head between the compartments, and the mill linings.

Ball Loadings

Balls used in the first compartment are larger than those in the second because much larger particles must be broken. Sizing distributions vary according to clinker characteristics and whether the mill is in open or closed circuit. Some clinkers contain small, hard nodules about 10–15 mm in size that grind slowly and may enter the second compartment by flowing through the large grates in the center of the partition. They accumulate at the discharge end of the mill, and at times a few large balls are added to the second compartment in an attempt to fix the problem. Table 7.3 gives the loading and size distributions of balls for some two-compartment mills.

Mill Division Heads

Although dry tube mills seem to be simple devices, in fact they are complex, and every aspect of their design and operation has to be considered carefully if good performance is to be achieved. For example, division heads are used for separating the grinding chambers that operate at different conditions (ball charge, ball size, linings) in the multiple-compartment ball mills. The aperture size of the diaphragm controls the top size of particles in the next chamber by restricting the passage of the grinding media and oversize material. Division heads can be classified under two major groups: (1) those used between drying and grinding chambers in mills for grinding raw materials and (2) those used between chambers in mills for grinding cement clinker (Figure 7.17).

Different designs are used for diaphragms between drying and grinding chambers and between grinding chambers. Scoops move material to the center of the division

TABLE 7.3 Loadings and size distributions of balls in two-compartment mills

Mill: 3.65 m in diameter by 10.3 m long, 1,300-kW motor, 49-tph feed rate Raymond separator: 5.5 m in diameter						
	Compartment 1			Compartment 2		
Ball size, mm	63.5	51	38	25.4	19	16
Load, ton	20.9	20.9	8.6	20.2	49.8	20.4

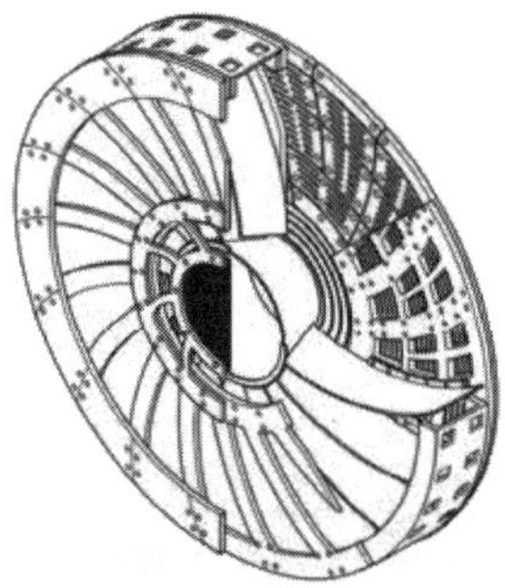

A

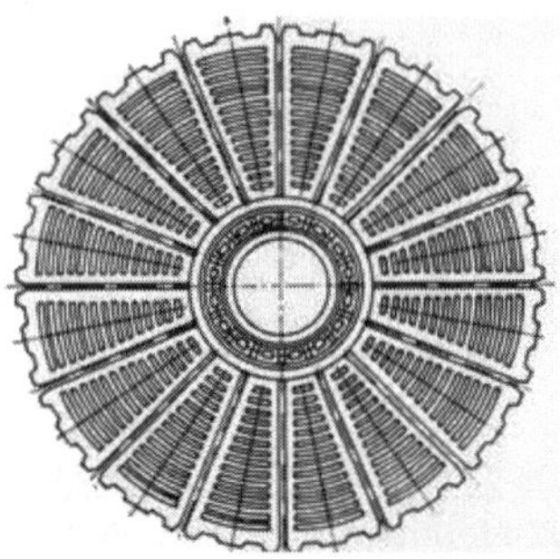

B

FIGURE 7.17 (a) Diaphragm between drying and grinding chambers (b) diaphragm between compartments in a cement clinker mill

head, from which it passes into the next grinding chamber of the mill. Division heads may have a coarse screen in the center so that the airflow is not impeded.

Some single-compartment wet-grinding ball mills also have discharge diaphragms and are called diaphragm or grate ball mills. Most larger-diameter wet-grinding ball mills in the mining industry are overflow mills that do not require grates. Problems resulting from grates becoming plugged with worn balls and the lower wear life of grates led to a general preference for overflow mills.

Classifying Liners

Figure 7.18 shows three types of classifying liners. As clinker grinding mills increased in length, it became necessary to devise a system by which the coarse balls were continually thrown back to where the coarsest particles of ore were at the feed end of the compartment. Hardinge had designed his "tricone" mill for this purpose, and later inventors sought to achieve the same result by modifying the design of liners.

Two Ball Mills in One Shell with an Intermediate Separator

Classification within grinding circuits improves efficiency and is used in the "double-rotator" two-compartment mill that Krupp Polysius introduced in 1951. This mill can be regarded as two mills in series in the same shell. Figure 7.19 depicts the mill, which uses one drying and two grinding compartments and a classifier.

The features of the double-rotator mill are

- The ore that has been ground in the first chamber is presented to a dynamic air classifier for separation of the fine particles.
- The rejects from the dynamic classifier are reground in the second grinding chamber and flow in the opposite direction through the mill.

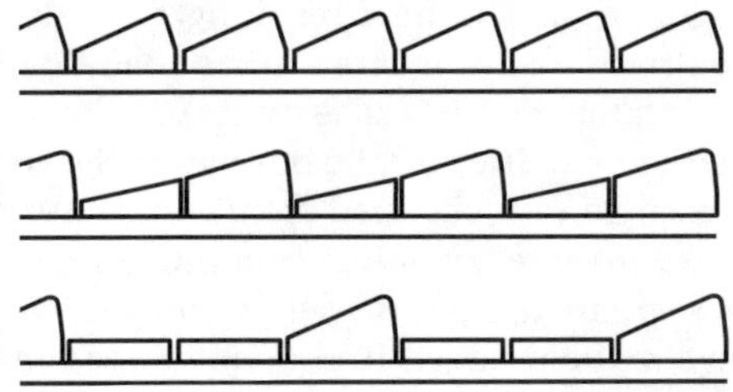

FIGURE 7.18 Classifying liners from (a) Carman (b) Slegten-Magotteaux (c) F.L. Smidth (Schiebe 1993; reprinted by permission of Klaus Graichen)

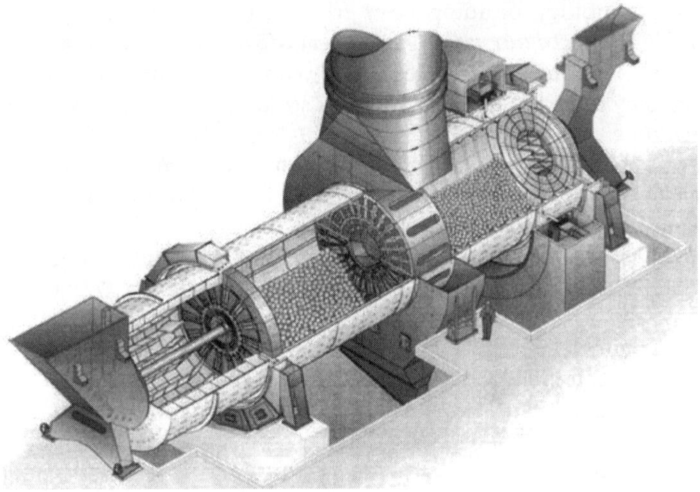

FIGURE 7.19 Krupp-Polysius two-compartment mill (reprinted by permission of Krupp Polysius)

The double-rotator mills built in the mid-1950s were 2.2–3.4 m in diameter and 6–8 m long, with 600-kW motors and feed rates of 20–60 tph. The largest mill now is 6.2 m in diameter, 25.5 m long, and is driven by an 11,200-kW motor. The need to grind refractory gold ores for roasting before leaching has revived interest in these mills. In 2000, two two-compartment double-rotator mills, 5.9 m in diameter by 21.3 m long with 7,500-kW motors, started working at Barrick Gold's Goldstrike mine in Nevada. Airflow to each mill was 110,000 m^3/min; the product from the first chamber was –6 mm, and the final product was 86% passing 74 µm. Approximately 20% of the dynamic classifier rejects went to the coarse end of the mill and 80% to the fine end.

The mills have proved to be suitable for dry grinding hard, abrasive ores for roasting and leaching at reasonable costs. Traditionally dry grinding has used more energy per ton than wet grinding for the same duty, so wet grinding has been preferred for metal ores. But the energy efficiency of dry-grinding tube mills is increasing, and they promise to be used more widely in the future with metallic mineral ores.

Grinding Coal in Ball Mills

Most of the grinding of coal for combustion burners is performed in roller mills, but air-swept dry-grinding ball mills operating in closed circuit with separators are also used. Coal is fed to the ball mill with heated air. The air is drawn through the mill by a fan located in the classifying section. The heat resulting from dry grinding and from the heated air is sufficient to drive off any moisture in the coal. Coal drying prevents coating in the mill, the classifier, and the air ducts. The oversize from the classifier is returned to the ball mill, and the fines are fed direct to the burners in the combustion unit.

The grindability of coal can be obtained from Bond (1952) or Hardgrove (1938) grindability tests. The Hardgrove test involves grinding 50 g of a prepared sample in a miniature pulverizer for a specified time and determining the amount of material that is retained on a 75-µm sieve in the product. It is a widely used test for coal, but the Bond test is preferred for most other minerals.

AUTOGENOUS MILLS

In this section, we make extensive use of Bond's review of early autogenous grinding systems, using his terminology for autogenous grinding (Bond 1964).

Pebble milling or ***secondary autogenous grinding*** is the process in which feed passing 19 mm or finer is ground by pebbles of ore with a maximum size of 76 mm or more that are extracted by screens from a coarse rock stream. This process was the first type of autogenous grinding used in plants.

Primary autogenous grinding is the process in which run-of-underground-mine ore or the product from primary crushers is the feed for mills. Both dry and wet primary autogenous grinding processes are used.

SAG grinding is the process in which 6%–12% of the primary mill volume is occupied by 100–150 mm (4–6 in.) in diameter steel balls to break the "critical-size" pebbles—the ones too large to be broken by the large rocks and too small to break finer particles. *SAG milling* has become a generic term to describe all types of primary mills that use rock or a mixture of rocks and balls to break rock.

The ABC (autogenous-ball-crusher) circuit is a particular type of SAG mill circuit that involves removing the critical-size particles from the SAG mill, screening them, crushing the screen oversize, returning the product to the SAG mill, and sending the screen undersize to the ball mill. Figure 7.20 shows a flow sheet for an ABC circuit.

Bond developed a media competency test which involved 100–160 mm rocks being rotated in a 1.93 × 0.32 m drum for a set time, sizing the product and testing the size fractions for grindability and resistance to impact. There is now little interest in this test, because media competency is not an issue with the SAG mills that are common today.

The first tumbling mills used hard pebbles from the beaches of France and Denmark on the North Sea to do the grinding. These pebbles were expensive, but the alternative grinding media was steel balls that were even more expensive, although they were heavier and broke particles faster. The term *autogenous* was apparently first used by Hardinge in 1940 during a discussion with mill workers about a mill he had designed that did not use steel balls (Robinson 1980). Autogenous grinding has since been defined as the "...grinding of ore by itself rather than by special grinding bodies distinct from the ore" (Bond 1964).

Autogenous grinding started in a small way at a gold mine on the Rand in South Africa in 1907. Tube mills used in the cement industry were being introduced then into South Africa to grind the stamp mill discharge finer for cyanidation, and a metallurgist decided to compare the grinding properties of pebbles made from the ore with pebbles from Denmark. The early tests showed that pebbles made from ore were successful, and autogenous grinding has been part of metallurgical practice on the Rand ever since. The process worked because the ore was a hard conglomerate cemented by silica and the conglomerate made ideal pebbles. The deposit itself was relatively homogeneous, which is always an advantage for autogenous grinding.

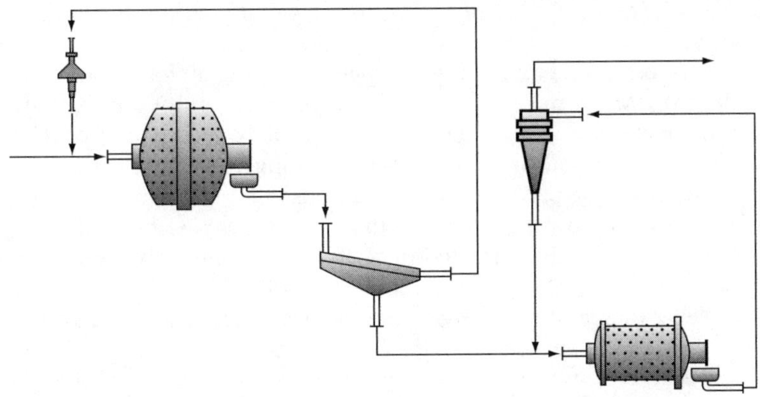

FIGURE 7.20 Flow sheet for an ABC circuit (JKMRC 2002; reprinted by permission from University of Queensland)

When tumbling mills were first used to grind gold ores in South Africa and Australia, the wear rate of the grinding media (the pebbles from the beaches of Denmark or France) was remarkably low at 2 kg/ton. But the cost of the pebbles was high at $5/ton, providing an incentive to develop pebble milling. In 1907, pebbles were used as grinding media on two occasions and these were the start of autogenous milling.

K.L. Graham at the Geldenhuis Deep Mine near Johannesburg ran a test using two mills 1.7 m in diameter and 7.1 m long to compare Danish pebbles and pieces of gold ore as grinding media. The 81-day test showed little variation between them. Engineers from other mines on the Rand verified the results in their own mills, and further experiments showed that 125-mm pebbles could be used to grind −0.5-in. feed. As a result, pebble milling became common practice in South Africa and was supported by development work. By 1932, SAG grinding was being used with up to 33% of 75-mm balls being added to the rock pebbles.

Hardinge, inventor of the conical mill and a later inventor and manufacturer of wet and dry autogenous mills, described a pebble mill at an AIME meeting:

> *(The feed) was 1 inch and finer. The mill contained 2000 pounds of 2 and 3 inch pebbles. The ore was crushed to 80 mesh size at the rate of 4 tons per hour at an expenditure of 17 hp. This type of mill appears to be specially adapted to the use of lumps of ore instead of the usual foreign bodies for grinding. (Hardinge 1955)*

> *The crops of birds filled with small pebbles, successfully break up whole grains preparatory to digestion. This is the natural prototype of the pebble mill. (Fischer 1944)*

Pebble milling was tested in North and Central America from 1912 to 1917 with results that were regarded as satisfactory at the time, but the idea did not catch on as it did in South Africa. Perhaps "the dam of professional hesitancy" (Robinson 1980) was a factor. Unlike South Africa, where many concentrators were clustered around Johannesburg, the concentrators in North and Central America were widely scattered and metallurgists had few opportunities to discuss ideas with their peers. Understandably, then, they were cautious about a major change that might go wrong and shut down a production system. But some early tests did give promising results, and it was established that pebble milling was successful for a variety of ore types. For some reason, however, the momentum was lost and little happened with wet pebble milling in North America for the next 30 years.

The difference between ball milling and pebble milling is that a ball mill with a ball charge weighing 4,700 kg/m^3 and occupying 40% of the mill volume draws approximately twice the power drawn by the same mill with a 40% charge of pebbles weighing 1,500–2,300 kg/m^3. The ball mill has a correspondingly higher capacity for the same reduction ratio. Converting pebble mills to ball mills will give higher plant capacity and requires only that the mill drive be changed to deliver the power needed.

Dry pebble milling was investigated in 1938 by the American Nepheline Company, which operated a dry 2.4 × 0.9 m conical mill to grind 4.5 tph of –7 mm syenite to –1 mm using 75-mm pieces of syenite as the grinding medium. The circuit was successful, and the use of dry grinding opened up questions about the merits of dry and wet grinding, gravity discharge, and air sweeping. These questions would only be answered by years of test work in plants.

Bunting Crocker, a metallurgist at Lake Shore Mines in Ontario, gave new impetus to pebble milling in 1948. The gold ore being processed at Lake Shore required grinding to 80% passing 30 µm, and steel balls (20 mm) were expensive. Crocker converted a mill (1.6 m × 5.2 m) to a pebble mill and operated it successfully. This led to the conversion of eight mills (2 m × 5 m) to pebble mills. He used 3–4 tpd of 75 × 50 mm lumps of screened ore as grinding media, and each mill ground 100–120 tpd. The rock media had a much lower cost than the steel balls, and the successful experiment created confidence in the use of pebble mills, particularly in the Canadian gold and uranium companies.

Pebble milling has been used continuously in South Africa from its introduction in 1907, with many improvements. By 1960, common practice was to use rod or ball mills as the first stage of grinding and pebble mills up to 3.9 m in diameter and 5.2 m long as the second stage. Metallurgists then became interested in the extension of the technology to primary autogenous grinding, which will be discussed in the next section.

In the 1960s in the United States, rod mill–pebble mill circuits were installed at the Anaconda Company's Kelly plant in Montana and Twin Buttes plant in Arizona. Pebble mills in the Kelly plant were 3.85 m in diameter by 6.45 m long, and in the Twin Buttes plant, they were 4.7 m in diameter by 9 m long. There were problems at each plant because the pebbles fed to the mill were not "competent" pebbles; that is, they were not satisfactory grinding media.

At one installation the nature of the ore was too fragile, the crushing plant did not produce enough pebbles, and the pebbles it did produce immediately broke in the 3.85-m (12.5-ft) diameter mills.

At the other installation pilot plant testing had shown that the ore made good pebbles. The crushing plant produced an adequate amount of properly sized pebbles. The pebbles were screened out of the ore in the crushing section of the plant and were then conveyed to the pebble storage area where they were dropped onto the pebble stockpile from a 19.4-m (60-ft) high stacking conveyor. There was breakage of the pebbles, and about 50% of the pebbles fed to the 4.7-m diameter mills were smaller than 25 mm and too small to be suitable media.

In the mid-1960s, the Duval Corporation installed four 6.45-m (20-ft) diameter mills in its copper ore concentrator near Kingman, Arizona. They were designed as intermediate autogenous mills to replace rod mills, and their feed was tertiary crusher product. The grinding media were 100-mm (+4-in.) products from a primary crusher but the circuit failed for lack of sufficient ore media, so it was converted to a conventional three-stage crushing circuit. Because of the structural design of the mill and the lack of power, the ball charge was limited to 29% of the mill volume. These were the first balls mills (6.45 m [20-ft] in diameter) used to grind copper ore. In 1967, an intermediate autogenous mill (2.3 m in diameter) was installed at the Western Mining Corporation nickel ore plant in Kambalda, western Australia, and it operated successfully for years, but interest in this form of autogenous grinding has faded.

PRIMARY MILLS

Wet Mills

In the early 1930s, the economic depression was in full force, and the only interest in new grinding machines was for milling gold ores, because gold was the only metal that was not falling in price. In 1933, gold rose from $21 to $35/oz. Alvah Hadsel, a mechanical engineer in California with considerable experience in the aggregate industry, devised a machine that he thought would replace an entire crushing-grinding circuit by lifting rocks high enough to cause them to shatter on impact when they were dropped onto a hard surface. In 1932, he built his first mill using this principle and installed it at the Beebe gold mine near Georgetown, California. It took the form of two parallel wheels (7.7 m × 1 m) that rotated at 2.66 rpm. Each wheel had 24 internal buckets (1.0 m × 0.6 m) to lift the ore and discharge it onto breaker plates (Hall 1935). This duplex mill, shown in Figure 7.21, was driven by a 74.6-kW motor. Water was added to the mill, and a 90-degree arc of each wheel was immersed in a concrete classifier tank. Fine particles overflowed the top of the tank, and the wheel collected the coarse particles for further breakage. The mill was reported to break 308 tpd from 0.3 m to 65% passing 75 µm.

By 1934, Hardinge had bought into Hadsel's enterprise, and mills were being sold as Hardinge–Hadsel mills. The largest of these was 8 m in diameter and 1.45 m long, and it ground 210 tpd of passing 203-mm tough silicified schist to 96% passing 150 µm using a 74.6-kW motor (Hall 1935). The breakage mechanisms in the Hardinge–Hadsel mill were shattering and, to a lesser extent, autogenous grinding. But many of the rocks or their fragments had to be dropped up to 70 times to be broken completely, and this caused excessive wear on the buckets. The industry verdict was that it was suitable for softer ores but not for hard, abrasive ores. In the mid-1930s, Hardinge redesigned it, building a high-diameter, short-length mill with a high peripheral speed that was called the Hardinge Cascade mill. A number of installations were successful, although critical-size particles started to become a problem.

In one sense the Hadsel mill may be viewed as a failure because the concept of lifting a rock to a height at which it would be disintegrated by its own weight when dropped onto a steel plate was unsuccessful. In another sense, it was a success, because it led to the concept that using large rocks as grinding media in tumbling mills was an extension of secondary autogenous grinding. Fifty years later, large-diameter SAG mills with length:diameter ratios in the general area of 0.5–1.0 were in use and are still being used today to grind huge volumes of ore at low cost.

By 1935, the partnership between Hadsel and Hardinge had ended, and Hadsel turned his attention to dry autogenous grinding. Once more he established a process that has become very important. Hardinge continued to work on the Cascade mill, and one of his more successful inventions was the "electric ear," which detected the change in sound emitted by the mill as the degree of filling rose or fell in response to ore change. The signal was used to maintain optimum grinding conditions automatically by controlling the feed rate. This was the start of feedback control for grinding mills.

Success for the Cascade mill came in 1958 after Hardinge presented a paper on autogenous grinding at the American Mining Congress. The paper's presentation led to 12 wet mills (5.8 m × 1.6 m) being installed in an iron ore concentrator built by the Pickands-Mather Company in Wabush, Quebec, in 1959 (see Figure 7.22).

A

B

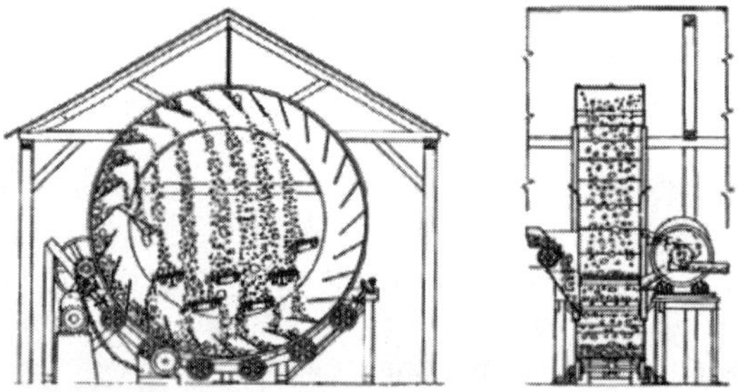

FIGURE 7.21 Hadsell mill: (a) original mill at the Beebe mine in California (Robinson 1980; reprinted by permission from CIM Publications) (b) how the mill was expected to work (Hardinge 1955; reprinted by permission from Primedia)

FIGURE 7.22 Bank of twelve 5.8 m by 1.6 m (18 ft by 5 ft) wet autogenous Hardinge Cascade mills installed in an iron ore concentrator in Quebec in 1959 (Robinson 1980; reprinted by permission from CIM Publications)

The installation was successful and must have done much toward overcoming the prevailing skepticism about the value of autogenous mills. Fortune favored Hardinge on that occasion if not on others, because the low-grade iron ores in the Quebec Labrador trough had well-defined iron mineral granular structures and were excellent feed for fully autogenous mills.

In 1956, primary autogenous grinding was investigated at the Grootlvei Proprietary Mine in South Africa. A pebble mill (3.9 m × 5.2 m) was converted to receive run-of-mine ore, and a Williamson controller was used to ensure that a constant load of ore was maintained in the mill. Jack Williamson was a Canadian engineer who was in charge of the Union Corporation research laboratory in Springs, South Africa, in 1948. He developed a pebble feed controller that became indispensable to the operation of primary autogenous grinding when this process started at Union Corporation in 1958. The basis of the controller was that the load of ore in an autogenous mill changed as the feed rate, size distribution, or hardness of the ore changed, and the power used by the mill rose and fell with the load. The Williamson controller measured the power and automatically adjusted the feed rate to maintain the mill load and power constant. After 55 days of plant operation, the controller was deemed to be a success and it was built into control systems for all Union Corporation autogenous mills. The principle used in this controller is still widely used for controlling autogenous mills although refinements have been added.

Dry Mills

With the success of primary autogenous milling and feed control at Union Corporation, primary autogenous mills became an accepted feature of Rand grinding practice. The autogenous mill circuits required about 25% more power than conventional circuits using pebble mills, but the smaller crushing circuit meant that there was lower capital investment and reduced maintenance cost.

Hadsel put his varied experience with the wet autogenous mill to good use when he built a dry primary autogenous mill in 1935. This large-diameter, short-length, air-swept mill initially rotated above the speed required to maintain a layer of centrifuged rock on the surface of the drum to reduce wear (the critical speed). In March 1936, a long mill (3.2 m × 1.6 m) was installed at the Harqua-Hala Gold Mines Company in Arizona, which eventually ground 90 tpd of –200 mm feed to 62% passing 75 µm. The mill eventually operated at 84% of the critical speed because of reduced capacity at higher speeds.

The new mill must have looked promising because, in 1937, Cominco, the large Canadian mining company, acquired the Canadian rights to the mill. It appointed David Weston, who had been mill superintendent at the New Golden Rose mine in northern Ontario, to manage the project. Cominco installed mills—one that was 3.9 m in diameter and 1.3 m long and another that was 2.9 in diameter and 2.0 m long—in its mines in Ontario and Northwest Territory, including the New Golden Rose mine, where it operated intermittently for 5 years, grinding 139 tpd to 66.5% passing 75 µm. Apparently most of the mills worked reasonably well, but in 1946 Cominco gave the rights to Weston, who set up Aerofall Mills Ltd. in Toronto. He made many improvements and by 1960 had sold about 20 mills. An Aerofall mill (6.0 m × 1.6 m) installed at the Benson iron-ore mine in New York in 1954 became "the first commercial autogenous mill in the world to process iron ore" (Roe 2000).

This sale was a milestone in autogenous milling because it opened up the low-grade iron ore market. Dry-grinding Aerofall mills were used in the third major installation of primary autogenous grinding mills in the Quebec Labrador trough, which was the Iron Ore Company of Canada's Carroll Lake plant. This installation and one for grinding asbestos ore were the high points and early successes of Aerofall mills. The patents granted to Weston for the dry-grinding Aerofall mills specifically covered mills with 6%

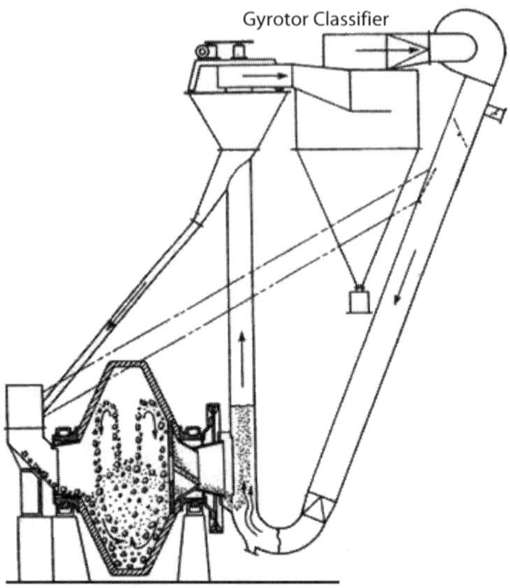

FIGURE 7.23 Dry mill and air separator (Hardinge 1955; reprinted by permission from Primedia)

of the mill volume containing steel balls and a mill speed of 82% of critical speed. The standard length for the mills designed by Aerofall Mills Ltd. was 1.9 m measured at the mill shell inside the liners. The Aerofall mill system was air swept and the air was heated sufficiently to dry any moisture in the mill feed. Some patents covered the installation of keying members at the throat of the mill at the feed and discharge ends to help lift the ore so that it would fall through the air stream. Figure 7.23 shows a dry mill and air separator.

In 1941, Hardinge supplied a dry mill (3.2 m × 1.6 m) to the Century Mining Corporation in Flin Flon, Manitoba. It had an electric ear for feed control and a 50-mm (2-in.) screen to separate the fine and coarse particles, allowing feed rates to the mill to be controlled. Although these features were important in the evolution of autogenous grinding and looked promising, they could not be thoroughly evaluated at the time because Century Mining Corp. closed during World War II and did not reopen.

The Growth of SAG Milling

By 1960, autogenous grinding had become reality. Aerofall Mills Ltd. and the Hardinge Company had sold 50 mills worldwide; Dorbyl in South Africa and Allis-Chalmers in the United States had started manufacturing them; and Morgardshammar in Sweden and Tyazhmash in Russia would begin manufacturing them soon. The road had been long and difficult, but the tenacity and resolve of Hadsel, Hardinge, Weston, and others had made wet and dry primary autogenous grinding possible and opened the door for the huge, low-cost-per-ton mills that were to come. No doubt there were dreams of the future during those hard years, but those dreams must have been scant recompense for the reality of heavy expense. Harlowe Hardinge Jr. (1955) made the wry comment that: "Our experience has now spanned a period of nearly one half a century and has been very costly to us to date to say the least."

The problems encountered by these pioneers were largely the result of the formation of critical-size particles that broke slowly and stopped the mill from grinding, because they filled it up. These particles resulted from the abrasion of larger rocks that

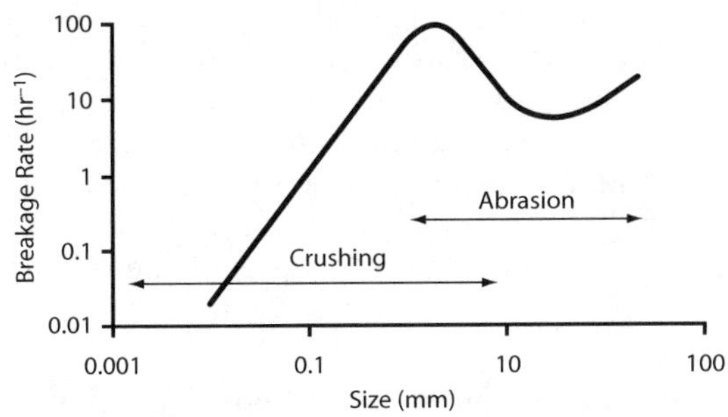

FIGURE 7.24 Critical-size problem in autogenous milling (Napier-Munn et al. 1996; reprinted by permission from University of Queensland)

were too small to continue to abrade quickly and too large to be crushed by the larger rocks. Figure 7.24 illustrates the critical-size problem.

The abrasion rate of particles in autogenous grinding decreases as size decreases. When particles are small enough, crushing starts and the crushing rate increases as size continues to decrease. There is a size range at which the total breakage rate is minimal, and if particles enter this size range at higher rates than they are broken out of it, the critical size builds up. The purpose of adding large steel balls to the mill is to break up an accumulation of critical-size ore. Primary autogenous mills were installed in the late 1960s in Ontario, Canada, and Tasmania, Australia, for grinding magnetite ores. By then it was clear that autogenous mills were controlled by the ore characteristics to a much greater extent than occurred when steel grinding media were used. C.A.R. recalls that autogenous grinding was referred to in Allis-Chalmers technical reports as "Grind me my way."

After World War II, the Mesabi Iron Ore Range in Minnesota in the United States produced many developments for crushing and grinding low-grade magnetic iron ores and pelletizing the magnetite concentrators. All the early concentrators used cone crushers and rod mills for preparing feed for coarse magnetic separation, and the concentrates were ground in ball mills for further magnetic separation, but one of the last three concentrators used wet primary autogenous mills with motors 11.6 m in diameter running at 9,000 kW. The other two of the last three concentrators used dry-grinding Aerofall mills. At one of these, a dry mill was rebuilt into a longer wet-grinding mill, and when that was successful all five mills at the other plant were rebuilt to longer wet-grinding mills, and five more wet-grinding mills of the expanded size were added. The original dry mills had ball charges so all ten mills were built with ball charges, and this plant became a large installation of SAG mills.

In the 1970s, the emphasis in primary autogenous milling was on iron ores, but work was also being done on the critical-size problems that were slowing down the use of autogenous milling for nonferrous ores. The principal difference between grinding ferrous and nonferrous ores was that the ferrous ore contained good-quality ore media, but the nonferrous ores generally required the addition of a 6%–12% ball charge containing steel grinding balls 75 mm and 100 mm in diameter to supplement the ore media. The ferrous ores would break along well-defined grain boundaries; the nonferrous ores with less well-defined grain boundaries tended to grind to coarser sizes than ferrous ores yet produced more fines. Primary autogenous mills with balls added to supplement the ore media were called SAG mills. Ultimately, the majority of installations were for grinding nonferrous ores and the majority of primary mills installed were called SAG mills without reference to whether or not they contained balls. The addition of balls added weight to the grinding charge and drew more power and, therefore, ground more ore.

The original concept was to use primary autogenous grinding mills to grind to the final product size. Even though this was successful, using primary mills to make feed for either pebble mills or ball mills was found to be more efficient. Studies on primary circuits showed that they could replace cone crushing and rod milling to prepare ball or pebble mill feed. Although the anticipated savings in the power needed for grinding did not materialize, considerable savings in capital and operating costs were realized by installing primary SAG mills.

It was also found that some of the ore in the mill was too small to be grinding media and too hard to be broken by the large ore media. This material, called pebbles, would accumulate in the mill, occupying valuable space and reducing potential capacity. If the secondary grinding was in pebble mills, the pebbles in the primary mill could be extracted and used as grinding media in the pebble mills. The pebbles in primary mills are referred to as the critical size. Removing the critical-size material from the mill, then crushing it in cone crushers and returning it to the primary mill or feeding it to the secondary mill increased circuit capacity. Many autogenous circuits now have pebble crushers in the circuit. Thus the use of cone crushers in crushing and grinding circuits has not been completely eliminated.

Primary dry-grinding autogenous systems had limited success in the field of extractive metallurgy. They were not energy efficient and had high operating and maintenance costs.

The diameter of the first practical size for primary autogenous mills was about 5.5 m. At the start of the 21st century, primary SAG mills 12 m and larger in diameter with motors larger than 30 MW to operate with ball mills 7.3-m in diameter and to operate as secondary mills with 12 m were also being discussed and planned.

By 1980, the expansion of iron ore mining had come to an abrupt standstill. At the same time the need for primary autogenous grinding for copper ore increased. The price of gold became attractive and gold ore mining expanded rapidly, increasing the demand for primary autogenous mills to grind gold ores. The demand for SAG mills to grind nonferrous ores increased because of their advantages in terms of capital and operating costs, and their sizes systematically increased even though their power consumption was up to 25% greater than that of crushers and rod mills. Figure 7.25 shows how the maximum sizes of motors in use increased during the period from 1958 to 1998.

For primary SAG mills it has been established that the mill operates best with the load in the mill occupying 25%–30% of the mill volume. In very large primary SAG mills, the feed opening has to be large enough to pass more than 1,000 tph of mill feed to prevent plugging of the feeder with large rocks. In both primary and secondary autogenous mills used for wet grinding, low-level diaphragms are used to prevent worn media and ore chips from floating into the discharge on the slurry, which would happen with an overflow discharge. The discharge grates must have wide, tapered slots so that they will not plug. The slots are wider in primary SAG mills than in grates for grate discharge ball mills. Avoiding problems with plugged grates is the primary reason that large-diameter ball mills have overflow discharges.

Figure 7.26 shows the inside of a large autogenous mill with its heavy liners. It illustrates the fact that the increase in mill size has caused problems in operation and maintenance, because parts are now much larger and heavier and jobs that previously had been done by hand now require machines. An example is the mechanical liner handlers which are now used to assist with replacing worn liners and installing new liners. A pebble extractor has also been invented by the Boliden Company in Sweden to remove pebbles from primary SAG mills and feed them into secondary autogenous mills.

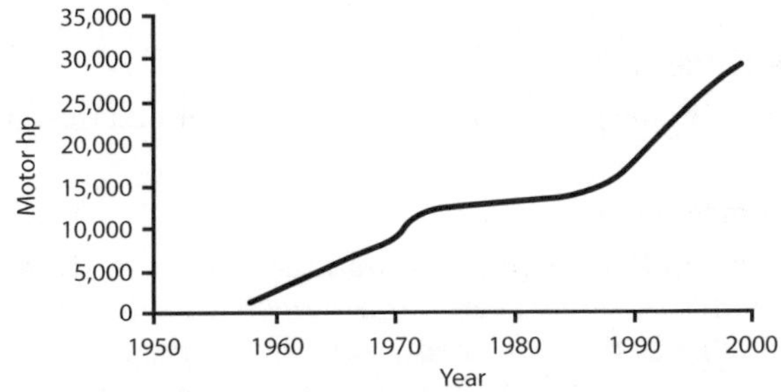

FIGURE 7.25 Increase in the maximum size of motors being used for SAG mills during 1958–1998

FIGURE 7.26 Mill end liners and mill load in a large SAG mill. The liners inside these mills require mechanical liner handlers for installation and removal (Tiggesbaumker 2001; reprinted by permission from Krupp Polysius).

A highlight of SAG milling has been the collaboration between grinding mill manufacturers and mining companies. Experimental work on very large mills is difficult, expensive, and potentially catastrophic if it goes wrong, yet the work must be done if the mills are to be accepted for use. The results of the collaboration have been very successful, as can be seen by the large mills in use and the high treatment rates obtained at low cost per ton.

TUMBLING MILLS: EVOLUTION OF DESIGN AND MANUFACTURE

Tumbling mills are basically described by the type of grinding media used, whether they grind the feed wet or dry, by the type of discharge arrangement, and the shape of the cylindrical part of the mill. All tumbling mills have common features:

- Drives
- Support bearings in which they rotate

- Cylinders
- Feed and discharge ends
- Wearing parts to protect the surfaces of the cylinder and the ends that are subject to wear
- Discharge arrangements

All of the tumbling mill manufacturers continue to have the same basic design challenges. Drives start with the energy source and end with the delivery of the energy to the cylinder. For tumbling mills the first energy source was the steam engine, followed by the electric motor. In some cases, internal combustion engines are used to drive tumbling mills.

The early mills and the smaller mills were driven by flat belts connecting the motor and the mill pinion shaft. The limitation on this drive was the ability of the belts to transmit the power. This limitation was solved by using V-belt drives, which were suitable to about 224 kW, but for more power the drives became so wide that they were difficult to install. This led to the use of direct-coupled synchronous motors, which opened the door to much larger mills.

Gears were a very important factor in mill design. The original gears were spur gears, which allowed single-tooth contact only and limited the power that could be transmitted. Higher power transmission was obtained by using herringbone gears, which allowed more than one-tooth contact. The gear teeth for each side of a herringbone gear are cut from the outside of the gear face; one side is cut, then the gear is turned over and the other side is cut. Manufacturing limitations caused the machining of the apex at the center of the herringbone gears to waver from being a true circle. The wandering apex, which was at the center of the gear, and the pinion required that herringbone pinions be free to move laterally in either direction. With the introduction of spherical roller pinion shaft bearings, the necessary clearance in the bearings was lacking for the free movement of herringbone pinions. Single helical pinions need to be held in a fixed position, so they did not need the same freedom as herringbone pinions. Consequently, single helical gears and pinion drives were used in grinding mill drives, instead of herringbone gears and pinions, where spherical roller pinion shaft bearings had been used. For mills drawing in the range of 5,000–9,000 kW (7,000–12,000 hp), single helical drives using two or more pinions to drive one large-diameter gear were developed, and the gear manufacturers built new larger-diameter gear-cutting facilities. Large-speed reducers were often installed between the motors and the pinions, with one motor and drive shaft for each pinion.

In the 1950s, the grinding mill manufacturers were building mills that required single helical gears with a diameter that approached the limit of what could be supplied by gear manufacturers. The availability of large diameter gears was more critical in Europe than in the United States. The first solution was the development of large-speed reducers (gear boxes) that could be coupled to the discharge trunnion of the grinding mill. The output speed of the reducer was the same as the mill speed. The speed reducers had either double or triple reduction of input speed. F.L. Smidth and Allis-Chalmers designed and built these drives, which were principally used in the cement industry. The sale of trunnion drives for tumbling mills ended when the gear manufacturers installed larger-diameter gear-cutting facilities.

Motors A second solution to the problem of ever-larger, single helical gears was developed in Germany, and it became useful when the problem occurred again in the 1970s. In the 1960s, Brown-Bovari and Siemens in Germany and General Electric in the United States developed the wraparound motor drives for long-length, two-compartment ball mills that would draw more than 2,240 kW (3,000 hp). These were for dry mills for grinding cement clinker in Germany. The design of these motors was the base from which the wraparound drives for SAG mills and single- or two-compartment large ball mills were developed.

The motor was an alternating current synchronous motor, and the sequence in the electrical starting and control circuit was to rectify the alternating current to direct current then back to a low-frequency alternating current to give the motor the same speed selected for the operation of the mill. The selection of the operating frequency for the low-frequency alternating current is a control variable that can be used to adjust the mill speed for changing grinding conditions. The design of large-diameter, high-horsepower SAG mills required drives that could deliver up to 22,500 kW, and wraparound motors with the motor rotors mounted on the shells of the mills were built for mill drives exceeding 9,000 kW.

Trunnion Bearings Manual grease lubrication was used on trunnion bearings. This put a limitation on the size of trunnion bearing with the maximum size being 0.87 m in diameter. Grease lubrication also lowered the allowable loading of trunnion bearings. With the development of self-lubricating oil bearings and then oil lubricating systems, much larger trunnion bearings were available as grinding mills became larger in size.

Primary autogenous mills built in South Africa and by Aerofall Mills were mounted on pad bearings around the mill shell. The South Africans needed this design for their high-speed mills with peripheral discharges. Aerofall's engineers wanted large-diameter trunnion bearings for its air-swept mills.

Structural Design of Mills

Mills and the materials used to build mills have to withstand the high stresses involved in each revolution of the mill with a complete reversal from maximum compression to maximum tension and back again. Mill design required a complete understanding of the stresses involved in one rotation of the mill. As the size of tumbling mills grew, the weight of the load in the mill increased and the stress on the shell and ends (heads) increased. The rotating cylinder of a grinding mill is a balanced drum that lifts and tumbles an off-center load, where the center of gravity of the mill charge is not on the centerline of the cylinder. The weight of the charge causes a deflection of the mill cylinder. Beginning in the 1960, when electronics and digital computers were making their marks on industry, mill manufacturers were concerned with methods of measuring stresses and determining the limits of various materials of construction, and were using complex computer programs for designing mill shells and heads. X-ray methods were borrowed from physicists to look for internal defects in plates and castings.

In the 1960s, single helical drives using two or more pinions to drive one large-diameter gear were developed by the gear manufacturers for mills using 4,500–7,500 kW (6,000–10,000 hp), and they built new, larger-diameter gear-cutting facilities. Figure 7.27 shows a primary SAG mill (11 m in diameter) with a wraparound 10,000-kW motor. These motors are low-frequency synchronous motors with a controller that allows the frequency and speed to be varied, which is important in the process control system for adjusting for ore variations.

Late in the 20th century, Hardinge, Allis-Chalmers, Dominion Engineering, Marcy, and Kennedy Van Saun, all designers and manufacturers of tumbling mills, were consolidated into Svedala International's Grinding Division. Svedala International then became part of Metso Minerals. At the start of the 21st century, the principal suppliers of primary autogenous mills are Metso Minerals, Dorbyl, Krupp Polysius (which purchased Aerofall Mills in the 1970s), the Fuller Company, and Tyazhmash.

FIGURE 7.27 Mill 11 m (34 ft) in diameter with a wraparound motor (courtesy of Metso Minerals)

MATERIALS FOR WEARING SURFACES

The working surfaces of crushers and grinding mills are lined with wear-resistant materials to protect the permanent parts. The history of wear protection is an integral part of the history of comminution, because it is a major factor in the operating costs for circuits. Replaceable surfaces were used as early as the 1880s in South Africa when cast-wearing parts made with Hadfield manganese steel were placed in crushers and mills. It was a good choice, because, when manganese steel castings are exposed to continuous impacting, they grow sideways as they wear. Because of this characteristic of manganese, it has been possible to design the castings and the crushing chambers to allow for balanced wear and growth of the castings. When the wear rate of the castings is lower than the rate of their growth, the intermittent operation of primary crushers allows time to remove the growth. Consequently, replaceable manganese steel castings have proved to be satisfactory wearing surfaces for jaw and gyratory crushers.

Materials used in mill liners now include impact-resistant alloyed cast iron, abrasion- and impact-resistant alloyed cast steel, rolled alloy steel, and—since the 1960s—rubber. Where heavy impacting occurs, chrome-molybdenum alloyed steel castings are used. Iron castings containing nickel resist abrasive wear but are brittle and break under impact. Nickel-iron cast works well as shell liners in rod mills but cannot be used in ball mills when the largest balls exceed 51 mm, because rods have a line contact with shell liners and a lower impact pressure than balls that have a point contact with the mill liners. Impact-resistant shell liners made of nickel-hardened cast irons have long lives in rod mills, but the end (head) liners, which are subjected to pounding by the lateral movement of rods, are made from chrome-molybdenum steel. Cast or rolled wear-resistant alloyed steel liners are used in ball and SAG mills.The story in tumbling mills was different. Manganese steel liners worked well in pebble mills and in small-diameter rod and ball mills, but when the rod-and-ball-mill diameters exceeded 2.25 m and larger grinding media were used, the heavier impacts caused the liners to grow faster than they wore. To protect the surfaces of the rotating parts of grinding mills, clearance between liners was minimal; as a result, growth in the liners without room to expand put forces on the mill heads and shells that caused cracking. If the edges of the liners were beveled, they would climb over adjacent liners as they grew, which would cause the liner bolts to break. When grinding mill diameters reached 3.55 m, the use of cast manganese steel liners was discontinued.

In the 1960s, Swedish and American companies used rubber for mill liners, which had the advantages of being much lighter in weight, easier to install, and less noisy. The problem was that when flotation reagents, such as xanthates, were added to the mills the rubber in the liners was degraded and softened. Broken rods cut rubber liners installed in rod mills. Due to the heat in dry grinding mills, rubber liners were not used in dry grinding mills.

Rod, ball, and SAG mill shell liners are made with either waves cast into them that act as lifters or with separate lifters. These lift the outer rows of grinding media to a higher position in the mill rotation. A rule of thumb is that the number of lifters in a ball mill should be twice the diameter in feet plus two, and in a rod mill it should be twice the diameter in feet. Liner wear in grinding shells affects mill performance, because it increases the internal diameter and volume of the mill, which results in more media in the mill, higher power draw, and higher mill capacity.

CONCLUSION

The development of tumbling mills to date has been the most significant stage in the history of size reduction. The Industrial Revolution not only developed a rapidly increasing demand for size reduction, it led the way to the development of crushers and tumbling mills that rapidly increased in size and ability to produce the amount of materials needed. The roller mills discussed in Chapter 6 and the tumbling mills covered in this chapter efficiently grind materials to as fine as 20 µm. In Chapter 8, we describe grinding processes and machinery that were developed to more efficiently grind materials to finer than 20 µm, even to 1 µm. These systems were developed in response to the ever-increasing demand for finely ground materials.

CHAPTER 8

Fine-Grinding Mills

In those days all paints were ground in very small quantities by hand, and it was not unusual for an apprentice to work on a couple of ounces of paint for many months. In the age when few colours were known it was the colloidal grinding of paints that enabled the great artists of the Italian Renaissance to find expression and pass on their wonderful creations to prosperity. (Schotz 1931)

In the preparation of india ink by the Chinese, according to Wolfe, a mixture of glue, lampblack, egg white, cinnabar, and musk was beaten 30000 times in a mortar until the black was finely divided and deflocculated by intimate admixture with the protective colloids of the formula. (Fischer 1944)

For thousands of years the fine particles required for flour, pigments, ceramics, and minerals were produced in querns, saddlestones, and edge mills. During the 20th century, the uses of and requirements for ultrafine particles proliferated. In this chapter, we give a brief overview of how fine-grinding technology evolved, along with some of the different machines and processes in use.

By 1880, the Industrial Revolution was in full swing, and rapid growth was occurring in the cement, electricity, chemical, pharmaceutical, and other new industries that required fine particles. Better machines were required to produce them, and inventors applied their skills to the problem, patenting both low- and high-speed mills. The ball mills that were introduced about 1885 were low-speed mills suitable for grinding to about 100 μm but unsuitable for very fine grinding, because small particles were not broken efficiently by falling balls. Other methods for grinding with balls were devised, such as stirring the balls to create collisions. High-speed stirred mills use high energies per cubic meter of mill volume and give higher rates of breakage of small particles and higher rates of production of fine particles. They will be discussed in this chapter.

The nomenclature used in fine grinding is summarized as follows:

- Ball mills: Separate grinding media are used, for both continuous flow grinding and batch grinding.
- Stirred mills: The material being ground is kept in motion by stirring rather than tumbling.
- Vertical high-speed stirred ball mills: defined as peg mills; horizontal high-speed stirred ball mills are defined as disc mills.
- Colloid mills and pin mills: High-speed mills in which rotating pins pass each other with little clearance, creating intense turbulence and shear forces in the slurry close to the pins.
- Vibrating and nutating mills: Very small balls are kept in rapid motion in a confined volume by moving the mill shell at very high speed.

- Jet mills: Stationary containers into which very high speed jets of particles are introduced so that they collide and cause the particles to break autogenously by impact and abrasion.

In this discussion, no attempt is made to define a "fine particle"; this has changed with time, and the size of particle that is considered to be fine in one industry may be coarse in another. Fifty years ago, particles of metalliferous ores that were less than 20 μm were regarded as fine; now the size is 5 μm. Historically, mineral pigments such as iron oxides were fine when they were ground to 10 to 20 μm; now they are ground to 1 to 2 μm. Synthetic pigments such as titanium dioxide are ground to 0.2 μm.

One of the problems with fine grinding has been in measuring size distributions of very fine particles so that the process can be controlled. Sieves have been available down to about 40 μm in size, but this size was not satisfactory for portland cement when it became a popular building material about 100 years ago, because the quality of the cement depended on its size distribution, and two-thirds of cement is less than 45 μm. The link between size distribution and surface area led to the Blaine number, which is related to surface area, being used in the cement industry as a measure of the quality of cement in terms of particle size.

In other industries that require very fine particles, a fineness of grind gage has been developed to determine the size of the largest particle in a product. This gage is a piece of steel into which a wedge has been machined from the surface to a depth of 10–200 μm, depending on the required particle size. A sample of the ground particles mixed with a liquid is placed at the deep end of the wedge and drawn toward the surface end with a blade. The depth at which the largest particles break the surface can be seen and measured.

The gage commonly used today is the Hegman gage shown in Figure 8.1. The wedge depth for various industries is

- Paint and pharmaceuticals: 125 μm
- Peanut butter: 20 μm
- Chocolate: 105–185 μm
- Ink: 10–30 μm

Fine products from grinding mills often have tight specifications on the amount of contaminates, such as iron, that is allowed in the product. This is why the grinding media are often natural silica pebbles or pebbles made from porcelains, most often an alumina porcelain.

SLOW-SPEED MILLS

The first mills to produce fine particles at a rate of a few tons per hour were slow-speed ball mills and roller mills that were developed about 1880. By 1920, there was a strong demand for fine grinding as industry revived after World War I. By then the ball mill was so successful that for many years better fine grinding meant finding ways to improve conventional ball mills rather than inventing better machines. Finer particles could always be made by grinding for longer periods in ball mills, and although this increased the cost, it did not involve a risk. But even then some new ideas were being tested. Ball or pebble mills were used for either continuous flow grinding or batch grinding.

Batch mills were made with removable openings on the shell of the mill. With the cover removed and the mill shut down and the media in the mill, the material to be ground was put in the mill and the cover was put on the opening. The mill was run until the material was ground to the desired product size. When periodic shutdowns and sampling for size were necessary, the mill was shut down; the ground material and the

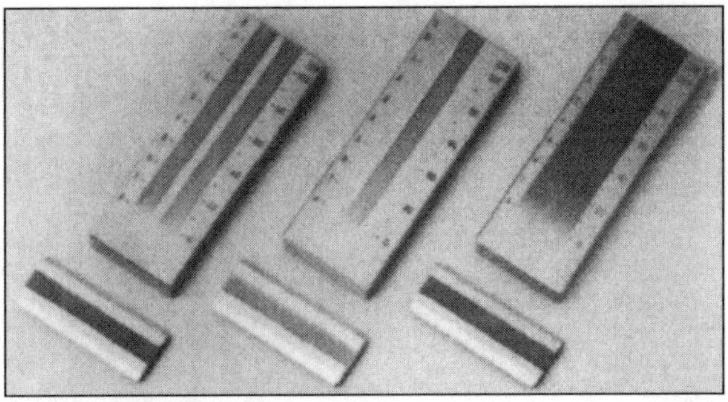

FIGURE 8.1 Two ink samples being leveled in gage paths by use of scrapper plate (Gardco 2004; reprinted by permission of Paul N. Gardner Co. Inc.)

media were removed and the media was separated from the ground material. When product purity required it, the mill and the media were washed before putting the next batch to be ground in the mill. Some of the batch pebble mill shells are made of porcelain. Batch grinding is used for paint and other fine iron-free products.

Ball mills, with lengths about twice the diameter, were designed to regrind rougher concentrates, middlings, and tailings when regrinding was needed. Stirred-energy mills, developed during and after World War II, particularly in Japan, proved to be more efficient than long-length ball mills for regrinding and are installed in sulfide ore concentrators for regrind services.

Vertical Stirred Ball Mills

> *As early as 1928 people started turning the hollow cylinder of the ball mill to a vertical position and started using a slow running agitator to keep the ball charge moving. The idea was to produce a uniform acceleration of the balls so that narrower particle size grinds results. (Netzsch 1974)*

In stirred mills the balls and solid particles are kept in a state of continuous movement by an internal stirrer rather than by being tumbled. An important use of stirred mills is to grind limestone to make a slurry to remove sulfur oxides from flue gases.

The advantages of vertical stirred mills are that balls do not have to be elevated and that ball-to-ball and ball-to-wall collisions are much more rapid than in tumbling mills. Attrition is the important breakage mechanism in stirred mills, whereas impact is significant in tumbling mills. The early stirred mills were low speed, and these are still used for grinding at the coarse end of the very fine-grinding spectrum. Higher-speed stirred mills were developed later and are now extensively used for very fine-grinding processes.

What appears to have been the first serious attempt to use a vertical stirred ball mill in the mineral industry occurred during World War II in Japan when Isokichi Iwasaki observed the operation of a horizontal ball mill while working in the aluminum industry and recognized that it could be used in a vertical form as a well-stirred chemical reactor (Nitta 1998). Some time after 1945, he was given financial support by Asahi Glass Co. to manufacture a prototype mill that was to introduce the concept to the Japanese mineral industry. The ability of the mill to produce ultrafine particles at low power consumption was recognized by Taketora Kobayashi of Nitchitsu Mining Co., which joined with Asahi Glass Co. to design, manufacture, and test a production mill in 1953.

The mill was successful, but patent trouble occurred between the partners, and the Nippon Milling Company was formed in 1955 to continue the work. The vertical mill

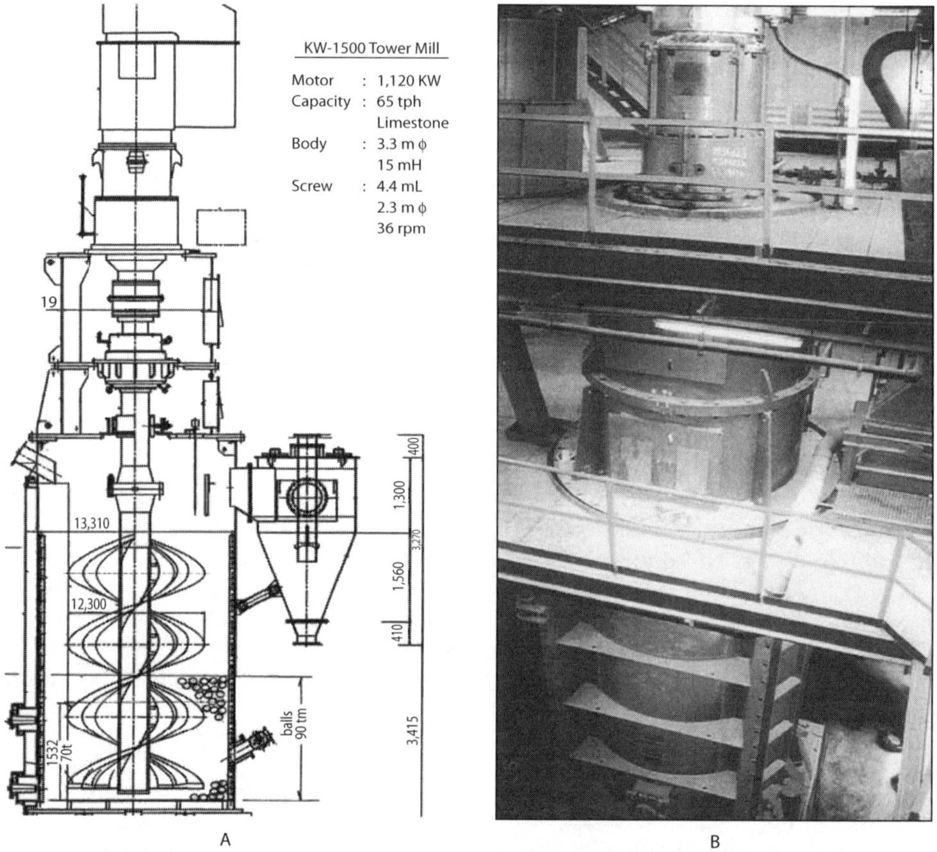

FIGURE 8.2 (a) Mechanism of operation of the Kubota stirred mill (b) Mill in plant is 3.3 m in diameter and 15 m high (Kubota 1998; reprinted by permission from Nippon-Eirich Company).

became known as the Tower mill (Figure 8.2), and in 1965 the company was renamed the Japan Tower Mill Company. Many of the early Tower mills were used in mines in Japan for fine wet-grinding purposes, and it gained wide application for the comminution of limestone for neutralization purposes, such as in flue gas desulfurization plants and for acidic waters. In 1980, wet Tower mills were installed in Cominco's lead/zinc and gold concentrators in Canada, and the mill was licensed to Koppers in the United States. Its main advantage in ore milling was that it could readily grind to about −25 μm and much finer if necessary. This was smaller than could be easily achieved by tumbling ball mills, which was a big advantage, but other benefits included low energy consumption, less noise, minimal vibration, and easy installation. By 1980, dry Tower mills were in use, and by 1990 ultrafine cement was being produced in them. In 1983, the Kubota Company, which had a large share of the market for grinding mills in Japan, acquired the Japan Tower Mill Company. By 1998, 242 mills had been sold in 15 countries, with the largest mill having a 1,120-kW motor and grinding 65 tph of limestone. In 1999, Nippon Eirich Co. Ltd. acquired the license for the Tower mill.

HIGH-SPEED COLLOID MILLS

The first mill for making particles about 1 μm in diameter was invented by Herman Plauson at his research institute in Hamburg in 1913 with the intention of making colloidal

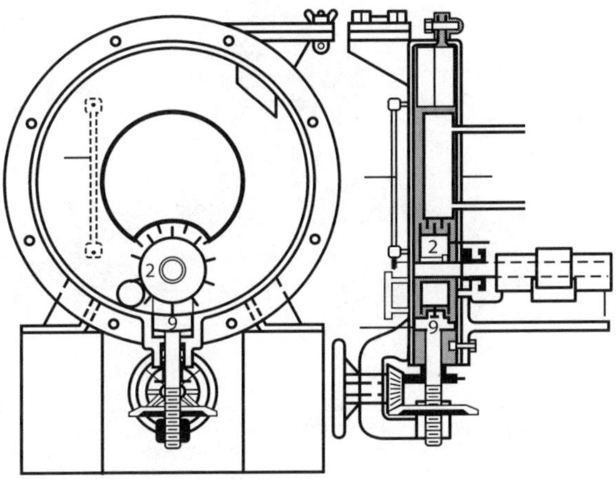

Sectional view of mill (front and side elevation)

FIGURE 8.3 Plauson-Block colloid mill (Forster and Reilly 1922)

coal, although it was 1922 before it was used in industry. Within a few years, colloid mills were being used for grinding pigments for inks and paints; phosphates; feldspars for use as fertilizers without sulfuric acid treatment; and sulfur for use in agriculture. The Plauson-Block mill is shown in Figure 8.3.

The mill consisted of a water-cooled casing and front cover, an axle carrying a set of beaters, and two sets of fixed anvils through which the blades of the beaters passed with 1–3 mm clearance. The axle in a 15-L mill was driven by a 22.4-kW motor and rotated at 3,000 rpm. The regime was intensely turbulent, and particles were shattered and abraded by impact.

> *It is stated that three 15-litre machines in series, running at 9000 rpm can grind 1000 kg of raw phosphate or 200 kg of cellulose in an hour to such a state of division that 85% of the particles are estimated to be of the order of .001 mm dia. (Forster and Reilly 1922)*

Another high-speed mill that was first used in the 1920s to make colloids is shown in Figure 8.4. "Machines of this type consist of a smooth moving surface which passes a smooth fixed surface at a high velocity (3 km per minute or higher) and with a small clearance (50 microns)" (Travis 1929). This type of mill was effective for deflocculation and dispersion and was used in the paint industry and for the breakage of mica and clays. Flat disks as well as cones were used as the grinding elements.

PIN MILLS

Medium-Speed Mills

A "soft size reduction" mill named the Triumph mill was made by Alpine in 1898 for preparing foodstuffs, animal fodder, and chemicals. The pin mill used the same principle as the colloid mill but with a lower speed of rotation. Figure 8.5 is the front page of the 1903 sales brochure.

This was the first of Alpine's pin mills, which were known as the Perplex line of mills. The 1903 Alpine patent described the mill "as a centrifugal mill with a multi-performance, stationary, grate-like perforated centrifugal ring located between the pin beaters arranged on a rotating beater disc" (Gelberg and Lange 1998).

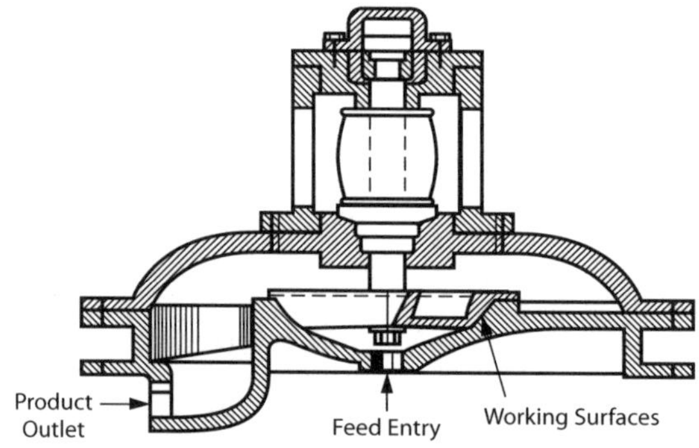

FIGURE 8.4 High-speed colloid mill (Travis 1929)

FIGURE 8.5 Alpine Triumph, an 1898 medium-speed size-reduction machine (Gelberg and Lange 1998; reprinted by permission from Hosokawa Alpine)

Alpine started as a small workshop in 1886 when an iron turner, Otto Holzhauer, began to make tools in a small mechanical workshop in Augsburg, Germany. He was said to have had an unbridled love of experimentation, and by 1900 he had a wide product range that included steam rollers, crushers, brackets for balloons, elevators, bottle filling and rinsing systems, and dryers. Financial problems arose in maintaining this wide product range, as a result of which Hugo Sachs became an investor in the company in 1902 and an owner in 1916. Holzhauer may have failed as a businessman, but he has a place in the history of size reduction because of his use of rotating pins and intense turbulence to disperse and break particles in a size-reduction machine. Alpine's Perplex mills were very successful with more than 17,000 being sold during the next 30 years. Size reduction in these mills involved two mechanisms: shredding and intense turbulence, which caused disintegration by dispersion. The original Triumph mill was designed to work as a shredder, but within a short time it was being used to grind materials that could either be shredded or shattered easily such as "cereals and grains of all types...and...a great variety of products such as ammonia salt, saltpetre, sulphur, cork, coffee, cocoa, glass, rubber, coal, leather waste, paper and sugar" (Gelberg and Lange 1998).

The pin beater mill made by Holzhauer in 1898 proved to be a versatile machine for general fine-grinding purposes, and they are now produced by many manufacturers. Some of the industries in which they are used are

- Chemical
- Pharmaceutical
- Animal feed
- Herb, spice, food, and confectionery
- Mineral powders, wood, and chipboard
- Plastics
- Ink and paint

The mills can be operated to produce various particle sizes, for example, less than 20 μm for pigments and pesticides to 150–500 μm for spices.

High-Speed Mills

Finer grinding is obtained in pin mills by using higher speeds, the product size controlled by the speed differential between the rotating discs that contain the pins. Figure 8.6 shows a mill with pin discs. The counter-rotating discs containing the pins in the housing and door are driven by separate motors. Table 8.1 gives the operating conditions for a large high-speed pin mill of the type shown in Figure 8.6.

BALL MILLS WITH FIXED SHELLS AND HIGH-SPEED STIRRERS

Vertical Peg Mills

After World War II, Netzsch-Feinmahltechnik GmbH, which had been established in Bavaria, Germany, in 1879 by two brothers to design and build machinery for the region's ceramic industry, took advantage of the rapid postwar recovery to diversify its product range and built processing machines such as filters, mills, and pumps. Its engineers had been investigating stirred mills for fine grinding for some years and had observed that the early vertical mills in which the grinding media were stirred at low speed had regions in which the energy intensity was low, resulting in some particles not being properly broken or dispersed.

In 1963, an agitator peg mill was patented by Willy John of Netzsch that was designed to create high and uniform energy intensity. It consisted of two concentric cylinders with pegs attached to the inner wall of the stationary outer cylinder and the outer surface of the rotating inner cylinder. The high peripheral speed of the inner cylinder and the interacting mixing patterns created by the short pegs provided a uniform energy intensity throughout the breakage zone, and the mill proved to be efficient for high-viscosity and high-solids content materials. It was particularly suitable for the grinding and dispersion of pigments in high-viscosity paint media.

For many years batch ball mills had been used for grinding paints, and long residence times were required. With much smaller balls moving at much higher velocities in the stirred mill, there was the same energy per impact but many more impacts so the work was done much faster. For paint manufacture, the ball sizes are of the order of 3–12 mm and stirrer speeds of the order of 70–400 rpm. High-speed peg mills are also used for mineral grinding, although the balls are larger, because the particles are hard and the products required are coarser. Figure 8.7 shows diagrams of the early "John" peg mill,

TABLE 8.1 Operating conditions for a large high-speed pin beater mill

Condition	Housing	Door
Disc diameter, mm	1,375	1,327
Speed, rpm	2,200	1,200
Motor, kW	200	132
Air volume, m^3/hr	17,300	

Source: Modified from CONDUX-Universal Mills brochure, authors' files.

FIGURE 8.6 Netzsch high-speed mill with pins in discs (Reprinted by permission from Netzsch-Feinmahltechnik)

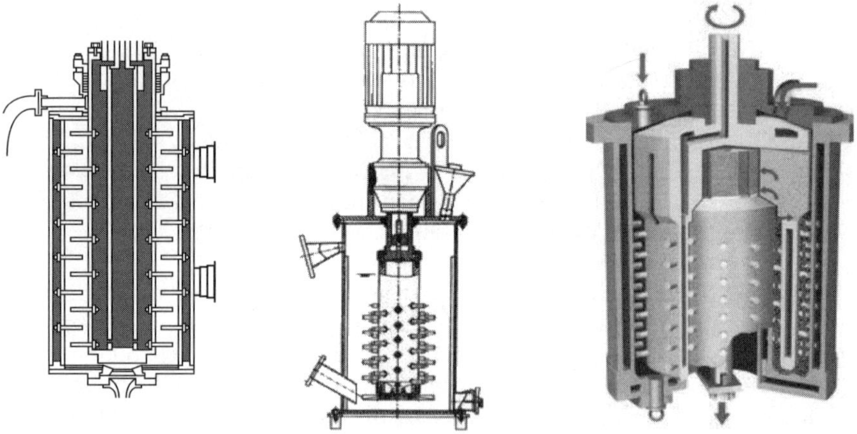

A B C

FIGURE 8.7 Three peg mills: (a) early Netzsch John mill (Netzsch 1974; reprinted by permission from Netzsch-Feinmahltechnik) (b) Sala-SAM mill for ore grinding (Denver Sala Basic 1994; reprinted by permission from Metso Minerals) (c) Drais Mannheim mill for ultrafine grinding (Drais Mannheim 2000; reprinted by permission from Buhler)

the higher-speed Svedala peg mill used for grinding minerals, and two very high-speed mills using very small grinding media for ultrafine grinding.

The media are glass, steel, or ceramic grinding beads in the range of 0.3–1.1 mm in diameter, and the product size is in the submicron range. The mill consists of three concentric cylinders, the inner and outer being fixed and the center rotating at high speed. The mixture containing beads, pigment, and fluid flows from the top down, through the outer high-intensity zone, then up through the inner zone and is centrifuged after the last stator peg is passed. The beads are returned through slots to join the incoming pigment and fluid. The ground product passes through a discharge screen and leaves the mill. This type of mill is used to manufacture liquid inks and toners, ultraviolet blockers, textile dyestuffs, paper-coating liquids, and other materials requiring ultrafine particles.

Horizontal Disk Mills

The improvement in grinding efficiency in a conventional stirred mill, as the media became smaller, led to sand with a particle size of 0.5–0.9 mm being used as the media and the shaft speed increasing from 1,700 to 2,200 rpm. This highlighted the importance of stirrer design to ensure that there was a uniform energy spectrum throughout the mill and that very high-viscosity fluids could be agitated properly. Netzsch developed an agitator that consisted of eccentrically mounted hollow discs on a rotating shaft that were staggered to form a screw that moved the media against the product flow. A typical use of these mills is for paint manufacture. The grinding beads can be glass, aluminum or zirconium oxide, or steel with a diameter of 0.5 mm and greater. The particles are usually ground to less than 5–10 μm. This type of mill has been adapted to grind fine-grained mineral ores to less than 7 μm. Figure 8.8 shows the IsaMill, an example of this type of mill.

The IsaMill uses 1–8-mm (0.04–0.31-in.) grinding media, which can be the ore itself, slag, sand, or steel or ceramic beads, and MIM Holdings Ltd. now uses thirteen 1.1-MW mills at three mines to produce particles that are −7 to −12 μm. The company also uses autogenous mills, tumbling ball mills, and Tower mills at three mines and is in a position to assess the link between power intensity, media size, and product size (see Table 8.2). Approximate product size of autogenous and ball mills is −50 μm and of fine-grinding mills is −5 μm.

BALL MILLS WITH SHELLS MOVING AT HIGH SPEEDS

Vibrating Mills

One method of keeping balls in motion so that they will collide and break particles trapped between them is vibration, and a vibrating ball mill was developed and patented in Germany in the 1930s. Vibrating ball mills are either vertical or horizontal and can be used as continuous flow or batch mills, and as individual units or in series. They are driven by mechanisms similar to those used for vibrating screens.

The grinding media in vibrating mills, which occupies about 80% of the working volume, can be steel balls, silica pebbles, or alumina porcelain pebbles. The grinding media move in a radial pattern, the distance of the movement being dependent on the throw of the mechanism. The speed of the vibrating mechanism is 1,200–1,800 rpm. At the end of World War II, Allis-Chalmers obtained some of the vibrating ball mill patents and found that the best applications for vibrating ball mill were for −5- to −10-μm products.

Vibrating ball mills are less expensive to install than tumbling ball mills drawing the same power and are smaller in size and weigh less. Essentially stationary units in that

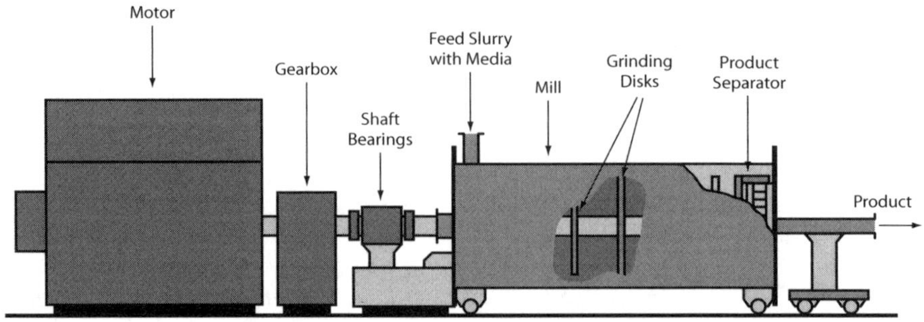

FIGURE 8.8 IsaMill, a horizontal disk mill used to grind fine-grained mineral ores (MIM Process Technologies 2002; reprinted by permission)

TABLE 8.2 Power intensity of tumbling and stirred ball mills

Mill Type	Mill Diameter, m	Mill Length, m	Installed Power, kW	Mill Volume, m³	Power Density, kW/m³	Media Size, mm
Autogenous	10	4.5	6,400	353	18	300
Ball	5	6.4	2,600	126	21	25–100
Regrind ball	3.2	4.8	740	39	19	~25
Tower	2.5	2.5	520	12	42	12–25
IsaMill			1,200	4	300	1–8

Source: MIM Process Technologies 2002.

they do not rotate, the mills do not need to be mounted in trunnion bearings. This eliminates the need for the massive foundations required by ball mills. The limited availability of large antifriction roller bearings for the vibrating mechanisms restricts the sizes of vibrating mills.

Centrifugal Mills

By 1970, mines near Johannesburg had reached a depth of 3,000 m, and consideration was given to underground milling to concentrate the gold minerals and to reduce the expense of transporting ore and tailings. This required mills that were much smaller than those currently in use, so the idea of centrifugal mills that would have higher breakage rates than tumbling mills was investigated. The concept had been proposed many years previously (Lloyd et al. 1982) but was not pursued on a large scale until the South African Chamber of Mines Research Organisation built and tested several centrifugal mills during the 1970s. Eventually, a 1-MW production mill, which was basically a horizontal grinding tube (1.0 m × 1.2 m) that gyrated at high speed in a vertical plane, was installed at the Western Deep Levels Mine near Carleton, Transvaal.

Philip Lloyd, leader of the research group, discussed the mill's performance:

> *The mill was held between two crankshafts with a throw of about 40% of the mill's diameter. ...Operated autogenously in open circuit at 230 rpm and fed with −100 mm quartzite, it ground 26 tph to 50% −75 micron at a power draw of 480 kW. Using 32-mm steel balls in closed circuit it ground 35 tph of −19 mm feed to 67% −75 micron at a power draw of 840 kW. Development was halted when it was found the circulating load could not be increased sufficiently. The throughput was limited at the discharge—the feed arrangements worked well and had a capacity*

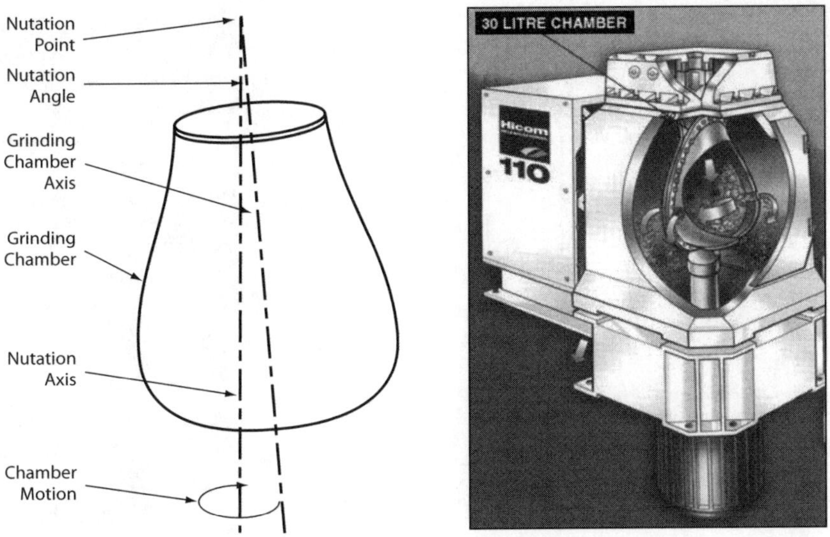

FIGURE 8.9 High-speed nutating action of the Hicom mill (Hicom International 2001; reprinted by permission)

> *of over 400 tph but although many discharge arrangements were tested we never managed to exceed a discharge much above 50 tph. Linings were simple and relining took place offline. The worn mill tube was lifted out of the cradle between the two cranks and a relined tube dropped in, a process that took about 15 minutes. (Lloyd 2001)*

Nutating Mills

Another method of keeping balls in motion at high speed is nutation, which is similar to the oscillation of a spinning top. The nutating action, shown in Figure 8.9, is used by the Hicom mill.

> *The grinding chamber is a truncated cone with a roughly hemispherical base, on a vertical axis of symmetry. The chamber axis "nutates" about a fixed point defined by the main bearing, the motion being similar to that of a conical pendulum in which the top is fixed and the bottom describes a circle. (Hicom International 2001)*

Hicom reports that the nutating mill has a power density of up to 2,500 kW/m^3 of mill volume and to give very high rates of breakage.

JET MILLS

> *The underlying principle of all jet action fluid energy mills is to induce particles to collide with each other at high velocity causing one or both to fracture into smaller particles. Multiple collisions enhance the reduction process and, therefore, multiple jet arrangements are normally incorporated in the mill design. (Grimshaw and Albus 1983)*

Jet milling is a low-volume process that has been used since the 1950s for making high-value, ultrafine materials. It is expensive compared with higher-volume processes so it is

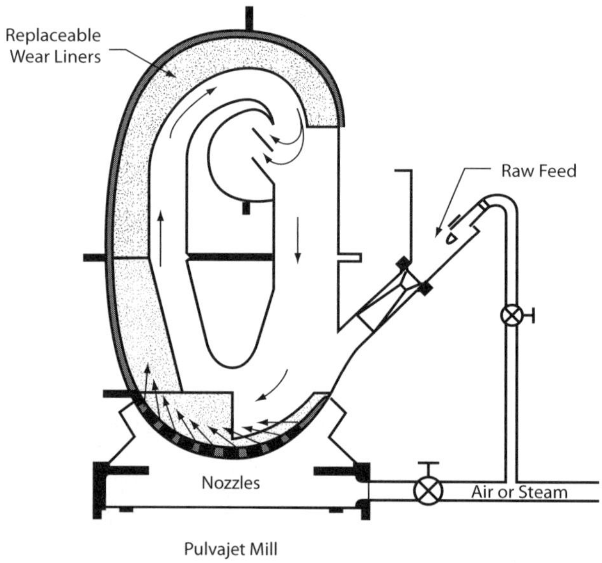

A

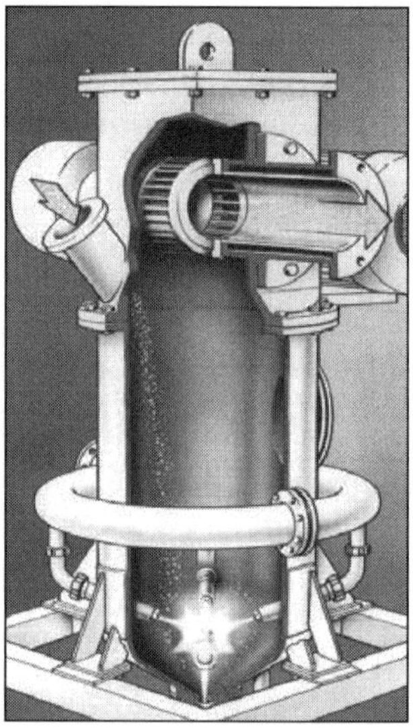

B

FIGURE 8.10 Jet mills: (a) vertical loop type (Grimshaw and Albus 1983) (b) fluidized-bed type (Hosokawa Alpine 1999; reprinted by permission)

only used for special applications. Some examples of products for which jet milling is suitable are

- Highly pure substances such as fluorescent powders and silica gels
- Abrasive materials such as tungsten or silicon carbide, and aluminum or zirconium oxide
- Powder coatings for thin film application
- Toners for photocopying machines and laser printers
- New ceramic materials
- Rare earths for the manufacture of magnets
- Contamination-free dental cement
- Fine industrial minerals that form flakes or needles
- Thermoplastics
- Temperature-sensitive substances such as waxes or resins

One type of jet mill is the older vertical loop mill shown in Figure 8.10a; another is the fluidized-bed jet mill shown in Figure 8.10b. In the vertical loop mill, the jet of fluid and particles enters at high velocity and passes a further series of jets that increase the probability of collision and breakage. Centrifugal action as the particles flow around the loop retains the coarse particles at the wall, and they reenter the breakage zone while the fine particles leave the unit through a center opening. In the fluidized-bed jet mill, high-velocity streams of particles suspended in a gas collide in a core breakage zone and, after leaving the core zone, form a bed that is fluidized by the gas. The fine particles are swept out of the bed by the gas and are classified in the rotating cage, the ultrafine particles leave the mill, and the coarser particles are returned to the fluidized bed and the breakage zone for further grinding.

The velocity of the jets is very high to ensure that the particles have sufficient momentum to break when they collide. The range of velocities through the nozzles is 500–1,000 m/sec and in the breakage zone is 200–400 m/sec.

> *The velocity of the driving fluid exiting the nozzles is directly proportional to the square root of the absolute temperature entering the nozzles, thus to obtain maximum energy from a given mass of fluid it is desirable to elevate the incoming temperature to the maximum tolerance temperature of the feed material. ...Optimum size reduction is achieved by the use of high pressure, superheated steam, generally in the range of 13.8–20.7 bars and 315–370 degrees C. (Grimshaw and Albus 1983)*

CONCLUSION

The demand for ultrafine products with even greater fineness will continue, and the value of these products justifies the use of individualized size-reduction processes and machinery.

This chapter completes the history of the development of methods and machinery for size reduction. Because it is often necessary to "classify" the discharge from the size-reduction machine–to obtain the desired results and control performance–we turn next to a brief historical review of size classification as used in size reduction.

CHAPTER 9

Classifiers

One of the early observations in size reduction was that a means was needed to separate a stream of particles into coarse and fine components. This was first done with flour; after the grain was ground in a manually operated grinding device, the flour was poured onto a cloth and shook by hand. The fine flour would flow through the cloth and the coarse particles would stay on the cloth. The fineness of the separation was established by the open spaces between the threads in the cloth. In the beginning, this was satisfactory for flour and other lightweight nonabrasive materials.

For aggregates and other rocks, size-separating surfaces were made with sticks, with uniform lateral spacing between the sticks for their full length. The bulk material to be separated was thrown onto the separating surface with shovels or dumped from buckets or similar vessels. The material was then mixed on the surface of the separator until all of the fine material had passed through the spacing between the sticks. It was soon found that hand shaking or pounding the frame holding the sticks would accelerate the speed of the separation, and then it was discovered that metal rods were more suitable for the separating surface than sticks. Metal plates with holes cut or punched in them made acceptable size-separation surfaces. Next, screening surfaces made with metal wire replaced the metal rods and plates, except for the grinding of large rocks, which damaged the thin wires. These led to the modern vibrating screens that are used in aggregate plants and in crushing circuits in mineral ore–processing plants. Screens make two-dimensional (width and thickness, but not length) size separations according to the size of the openings in the screening surface.

As the processing of minerals evolved, operators noted that flowing water and air would carry small particles, with coarser particles settling out. The faster the velocity, the larger the particles carried by the flow. This led to size separation based on the weight of the particles, which varies with the specific gravity of the material. From this, hydraulic classifiers were developed for wet grinding and air classifiers were developed for dry grinding. Both types are used for size separations that are finer than those obtained with vibrating screens.

HYDRAULIC CLASSIFIERS

> *Classifiers are devices for subjecting comminuted ore to the action of water either in such a way that a division of the ore particles is made into two or more products according to relative settling powers, or in such a way that all the ore material is settled together for dewatering purposes. ...Classifiers all have a carrying current of water...which carries forward all grains remaining in suspension. (Richards and Locke 1940)*

> *Classification equipment evolved from two sources, the simple expansion chamber and the Mumford and Moodie Separator. In the former, coarser particles drop out of an air stream as its velocity is decreased upon expanding to a larger space. Baffles, vanes or other directional and impact devices were later incorporated in the*

> *expansion chamber to change the direction and provide collision surfaces to knock out coarser particles. ...[In the Mumford and Moodie Separator] solids are fed into a rising air stream using a rotating distributor plate that imparts a centrifugal force. Coarser particles drop into the inner cone; the fines are swept upward by the action of an internal fan. (Klumpar, Currier, and Ring 1986)*

The abrasion of rocks from wind and water and the classification and transport of fine solid particles have been some of the processes that have shaped the surface of the earth for eons. But it was not until the late 19th century that controlled classification was used as an industrial process and is now used extensively. It is an integral part of wet- and dry-grinding processes for two reasons:

1. Every continuous grinding process has internal classification. Particles of different sizes leave mills at rates that depend on their size and specific gravity, the smaller and lighter-weight particles being discharged at higher rates. The effect of in-mill classification can be seen in the coarse rocks and pebbles contained in autogenous mills.
2. External classification is used to ensure that the products of grinding meet the required specifications. It is the self-standing external classifiers that will be discussed in this chapter.

In 1900, grinding in wet ball mills was becoming accepted as a viable method for producing fine particles, but the mills were operated in open circuit. This meant that the products contained too many very fine particles if all the particles larger than the required top size were to be eliminated or too many coarse particles if overgrinding was to be avoided. The dilemma could be solved by discharging particles from the ball mill before too much overgrinding occurred, classifying it, and returning the coarse particles to the mill for further grinding. Classification of slurries was not new. The types of classifiers in use before 1900 were those in which

- Rising currents of water sorted the particles into size fractions and removed the fine particles. Classification was precise but water use was large (Figure 9.1a).
- The slurry flowed horizontally across settling boxes of increasing size, with the coarsest particles settling in the first box and finer products in later boxes (Figure 9.1b). This is the Rittinger classifier.

These classifiers produced sized fractions of ore for concentration by jigs or tables and were also used to remove −200-µm particles of quartz or their equivalent when they interfered with concentration. The fine product was very dilute.

John Dorr transformed wet classification. His place in the history of engineering was established when he invented a machine in which mineral slurries could be classified into fine and coarse streams with the fine stream containing up to 50%–60% solids and particles up to 200–300 µm, depending on the specific gravity of the mineral. He was also able to change the size at which the split was made, which was a significant advance as far as controlling the process was concerned. His classifier improved the precision of grinding and changed it into a continuous, high-capacity process for producing particles broken to the size at which they could be concentrated efficiently.

The Dorr classifier used the force of gravity to separate heavy and light particles. It may seem surprising that cyclone-type devices were little used for the classification of slurries until the 1960s, although centrifugal forces generated by the rotating flows in cyclone-type devices had been used since 1885 to separate dry heavy and light particles suspended in air. Their capital cost was much lower than the cost of mechanical classifiers, probably because for many years the available slurry pumps were not good enough

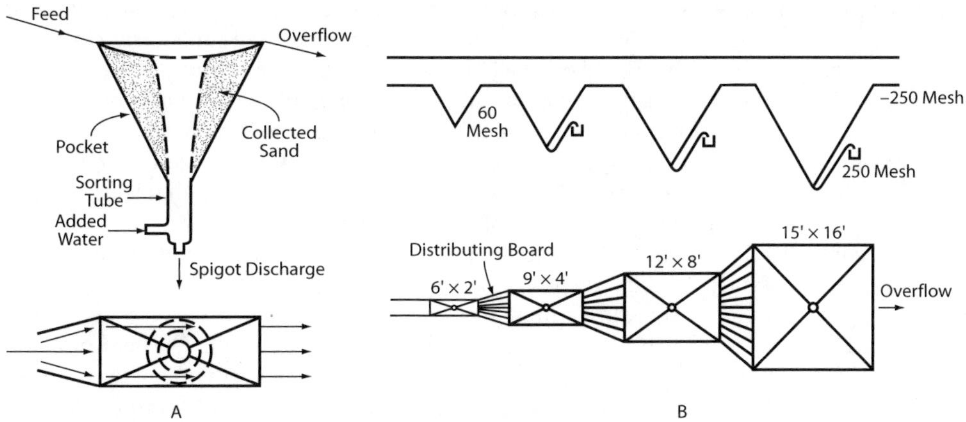

FIGURE 9.1 Classifiers: (a) hydraulic (b) surface (Taggart 1927)

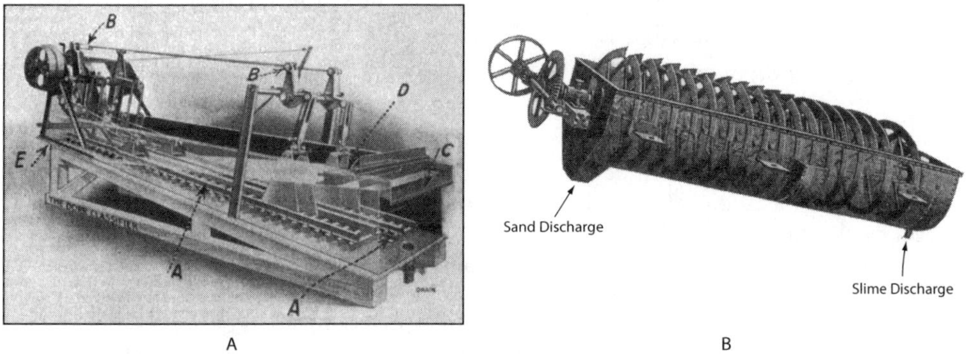

FIGURE 9.2 (a) Dorr rake classifier: A = rakes, B = rake mechanism, C = fines discharge, D = feed box, E = coarse discharge (Richards and Locke 1940) (b) Akins classifier in which a screw conveyor moved coarse particles to the discharge (Truscott 1923)

to generate the required pressures. But there was no alternative to centrifugal classification for dry particles, so cyclone-type air classifiers had to be made to work.

RAKE CLASSIFIERS

Figure 9.2a shows the Dorr classifier as it was in 1940, and it has changed little since then. Mineral companies quickly accepted the rake classifier and thickener, and closed-circuit wet grinding became part of mineral processing technology. With its reciprocating rakes, the Dorr classifier soon had competition from the Akins classifier (Figure 9.2b), which used a screw conveyor.

HYDROCYCLONES

For the 50 years leading up to 1955, the only serious competitor for the Dorr rake classifier was the Akins spiral classifier, which operated in a similar manner using gravity as the separating force but with coarse particles being removed by a rotating spiral rather than a rake. The settling pool was quieter in the spiral classifier, and the classification would have been marginally sharper, but the difference was not large and the choice of a classifier would have been made according to capital and maintenance costs rather than classification efficiency. By 1950, it had become evident that rake and spiral classifiers would be an impediment to the design of the high-throughput grinding circuits then required because of their high capital cost and limited capacity to transport coarse particles back to the mill. The demand for iron and copper minerals in particular was growing, and the new concentrators required high-capacity circuits. The idea of high-energy classification in cyclones using centrifugal force was well known, and reliable centrifugal slurry pumps were available that were capable of pumping high volumes of slurries at high pressures. The time for hydrocyclones had come, and they were about to change wet-grinding circuits from relatively low- to very high-capacity processes.

The inner and outer spiraling flows in hydrocyclones, which carry the fine and coarse particles into the product streams, are shown in Figure 9.3. The hydrocyclone may seem to be a very simple processing unit because it consists only of a few pipes and a cone, but its performance as a size separator is not easy to predict accurately. Two reasons are

1. It is influenced by both the size and specific gravity of the particles in the feed so that particles of equal size from different minerals will behave differently with a greater proportion of particles of the heavier mineral than of the lighter mineral entering the "coarse" product.
2. The fraction of ore in the feed that short circuits directly to the coarse product is difficult to predict.

Despite these problems, hydrocyclones are the classifiers most commonly used in wet mineral processing plants because they are inexpensive to install and operate, reasonably efficient as separators, and nothing goes wrong with them. How they have come to occupy their present dominant position in wet-grinding circuits will be discussed in this chapter. An offshoot of the hydrocyclone is the dense-medium cyclone in which finely divided material with a high specific gravity, such as magnetite or ferrosilicon, is added to the slurry. This increases the specific gravity of the liquid, and the dense-medium cyclone can be controlled so that lower specific gravity minerals float and higher specific gravity minerals sink. Dense-medium cyclones are used to separate coal from silica and silica from heavier metallic minerals such as hematite, galena, and sphalerite. In the rest of this section, for simplicity we refer to hydrocyclones as cyclones.

The Early Years: 1890–1950

The history of cyclones goes back to 1891 when E. Bretney obtained the first patent (Bretney 1891). The Bretney cyclone was designed with a closed apex for intermittent discharge and was the forerunner to present-day desanders that are used for separating sand from water in pressurized water systems. The Bretney cyclone had a roof entry feed inlet that canted down into the main cyclone body. This first cyclone was a crude, unlined, mild steel cyclone, but it paved the way for bigger and better things to come.

Between 1891 and 1939, a number of patents were granted on cyclones, but the record shows very few commercial installations. One of the earliest references to a commercial installation during this time was a cyclone 1.6 m in diameter that was installed in a U.S. phosphate plant in 1914. Most of the patents granted during this early period covered the use of cyclones for cleaning paper stock. The early designs were typically crude and centered on the use of smaller-diameter cyclones handling relatively low flow rates. Some commercial installations of pulp and paper cyclones started to appear toward the end of the 1930s. Most of the work on these cyclones was being carried out in Europe. Interestingly, one of the first references to a cyclone with a siphon on the overflow and an elutriation of water at the bottom was recorded in 1929 in an English patent.

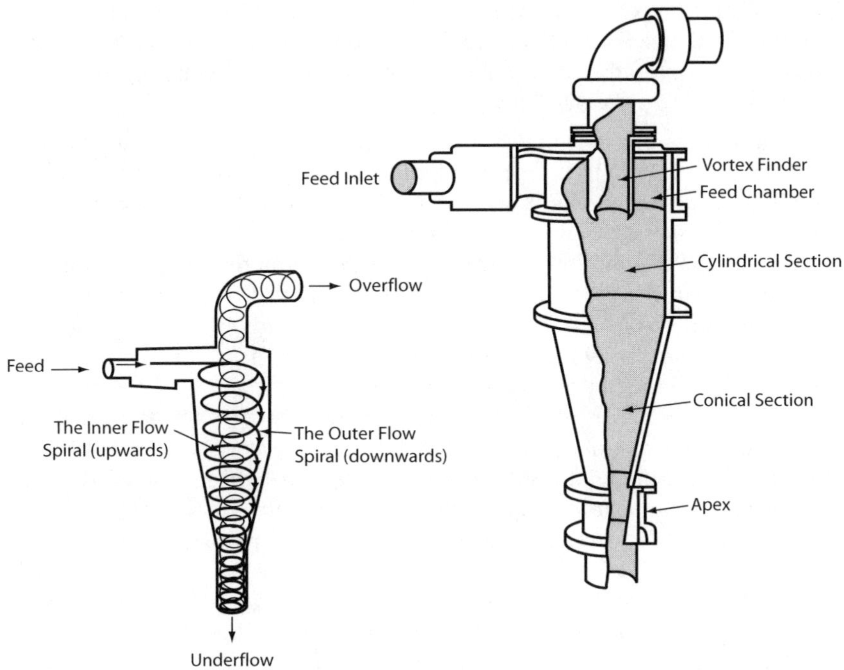

FIGURE 9.3 Hydrocyclone, showing the main components and principal flows (Napier-Munn et al. 1996; reprinted by permission from University of Queensland)

In 1939, Dutch State Mines (DSM) started investigating the use of cyclones for the cleaning of coal. The first use of cyclones by DSM was to dewater the sand used to make up suspensions for heavy-medium separators. Around the same time, the Powell Duffryn Company in the United Kingdom was looking at using cyclones for dewatering coal ahead of screens, and it obtained a patent for a cyclone in which the apex section could be replaced to overcome abrasion.

In 1943, the first work on the use of conventional cyclones for liquid–liquid separations was undertaken at the U.S. Atomic Energy Commission, where scientists tried without success to use the centrifugal forces in the spiraling flow in cyclones to separate isobutanol and water (Tepe and Woods 1943).

The most prolific period for the development of the cyclone occurred from 1939 to 1948, when M.G. Driessen of DSM led the effort to develop the use of cyclones in coal preparation. The main applications investigated during that time were the dewatering of the sand suspension and the actual cleaning of coal from sand using the sand suspension. According to legend, the cyclone's use for the cleaning of coal was due to a perceptive observer. Apparently, a cyclone that had been used by DSM staff for cleaning sand used in a coal concentration cone was dismantled for maintenance and it was noticed that the overflow pipe contained very clean fine coal. DSM engineers realized that two processes were involved:

- A dewatering process in which the sand was concentrated in the outer spiral and left the cyclone through the apex while most of the water left through the vortex finder

- A concentration process in which the centrifugal force enhanced the efficiency of the sand suspension as a dense-medium separator and caused the coal particles that contaminated the sand to move from the outer to the inner flow in the cyclone and enter the overflow

This observation about the effect of centrifugal force on gravity concentration led to the introduction of cyclone-based dense-medium equipment that changed the efficiency and economics of the beneficiation of coal, iron ore, and some sulfide ores, and to DSM developing a large business in dense-medium cyclones.

Driessen obtained many patents and published numerous papers, which fueled a worldwide interest in the commercialization of cyclones (Driessen 1948). Many people started using cyclones for clarifying water, removing solids from drilling mud, and for mineral concentration. In 1944, the Humphrey Investment Company, using Driessen's published works, developed a Humphrey centrifuge—the first large-diameter commercial cyclone with a vortex finder, used to dewater mineral slurries ahead of the Humphrey spiral. Ironically, the original DSM cyclones did not have vortex finders, and it was not until 1948 that the commercial DSM cyclones incorporated these. The DSM patents were commercialized by Stamicarbon N.V., which licensed the technology to companies such as the Dorr Company (which later became the Dorr Oliver Company) for minerals and Heyl Patterson for coal.

The Dorr Company started marketing the Dorrclone in 1948. Stamicarbon continued research on the cyclone and in 1948 filed for the first known patent on the use of conventional cyclones for separating liquids and also for a patent on the use of an adjustable elastomer apex. The year 1948 saw tremendous activity with the cyclone around the world:

- The Fuel Research Institute of South Africa started working with cyclones for coal washing in South Africa.
- F.T. Doughty published one of the earliest technical papers on cyclones for mining in *Mine & Quarry Engineering* entitled "The Cyclone—Its Use for Mineral Concentration" (Doughty 1948).
- One of the earliest reported uses of cyclones in the U.S. mining industry was small-diameter cyclones used at Chino Mines Division, Kennecott Copper, Hurley, New Mexico.

As the awareness of cyclones started to rise, other manufacturing companies besides the Dorr Company started looking at the cyclone as a viable product. In 1948, American Cyanamid started experimenting with a small-diameter DSM cyclone for heavy-medium coal separations and continued this work until 1950 when it turned over the designs and work to Kelly Krebs. Krebs had retired from American Cyanamid and started his own company, Equipment Engineers, in San Francisco, California. Krebs was acquainted with Bob Clarkson, who invented the Clarkson reagent feeder, and Krebs persuaded Cyanamid to license and manufacture it.

In 1948, Driessen left Europe and came to the United States to work for Heyl Patterson. The coal industry continued to develop uses for cyclones, and most of the patents in the late 1940s and early 1950s were related to coal. In 1949, however, other industries began to recognize the potential of cyclones. The first reported tests of small-diameter cyclones for starch separations were conducted in Holland, resulting in the first commercial installation for gluten–starch separation. The Dorr Company, in cooperation with DSM, developed the first TM (tandem-multi) 10-mm cyclones for starch separations. Other reported uses of cyclones in 1949 included phosphate processing in Florida and iron ore processing in Minnesota and Russia (Wright 1949). It was around this time that the Dorr Company started to push the use of the cyclone in the mining industry.

D.A. Dahlstrom was an early investigator of cyclones, first at Northwestern University (Evanston, Illinois) in the late 1940s, and then at the Dorr Oliver Company in the early 1950s. He published some of the early modeling work on cyclones and was one of the first investigators to try and find out how a cyclone worked (Dahlstrom 1952; Emmett and Dahlstrom 1953). Dahlstrom's work with the cyclone raised its visibility and led to significant research work at various universities around the world, including significant research by Kein and Ellefson at Northwestern on the use of cyclones for both liquid–solid separations and liquid–liquid separations. In 1952, Dahlstrom and the Dorr Company patented the first cyclone with a hydraulic water addition at the bottom to improve the efficiency of the cyclones.

The Rise to Dominance of the Cyclone: 1950–1970

Most of the work in the mining industry up to 1950 was for dewatering or desliming. The big breakthrough application for the cyclone, however, was just around the corner. In 1951, metallurgists at the San Manuel operation of Magma Copper Corporation (San Manuel, Arizona) experimented with a specially designed cyclone for closed-circuit grinding. The results were inconclusive, but people were starting to take notice.

Late in 1951, the Dorr Company installed a 0.65-m Dorrclone for closing the grinding circuit at Tennessee Copper in Copperhill, and other companies–including Climax Molybdenum and Anaconda Copper Company–started to test cyclones for this application. From 1951 to 1954, numerous researchers and companies around the world started to experiment with the use of cyclones for closed-circuit grinding. Early papers on mill-cyclone closed circuits were concerned with preparing gold ore for leaching in South Africa (Dennehy and de Kok 1953) and limestone for feed to cement kilns in the United States. E.B. Fitch and F.J. Fontein published many papers on the effects of design and operating variables on the performance and selection of cyclones (Fitch and Johnson 1953; Fontein 1961). Although the Dorr Company and other manufacturers pioneered the use of cyclones for closed-circuit grinding, it was Equipment Engineers (later Krebs Engineers) that eventually took the market away from Dorr Oliver.

In 1950, when Clarkson was mill superintendent of the Bradley Mining Co. in Idaho, Cyanamid abandoned the equipment business and returned the rights of the feeder to him. Krebs and Clarkson eventually formed the Clarkson Company to manufacture and sell the feeder, with Krebs responsible for sales in San Francisco and Clarkson for manufacturing, which was carried out at the Boise, Idaho, airport. Cyanamid also gave the Clarkson Company the rights to some design concepts for the cyclone that Krebs and Clarkson had been working on. Equipment Engineers took responsibility for cyclones and the Clarkson Company for feeders. In 1951, Equipment Engineers acquired the rights to a forced vortex cyclone, called the Centricone, from its inventor, Norris Goodwin (see Figure 9.4). The first customer was American Cyanamid, which used it to classify phosphate ores.

In 1953, wear problems with the impellor in the Centricone led to Equipment Engineers selling the rights to Oliver Filters, which was soon to become Dorr Oliver. Equipment Engineers then started work on a more conventional cyclone with an involute style feed (see Figure 9.5). This proved to be one of the features that made the Krebs cyclones manufactured by Equipment Engineers the industry standard for high-volume applications. The other feature was the thick molded-rubber linings. During this time, numerous studies were being performed using the cyclone in industries outside coal and mining, because it was recognized as a simple yet efficient separating device. Up to this point, most of the technical research on the cyclone was limited to plant applications and, with the exception of Dahlstrom and Driessen, very little had been done to examine how a cyclone actually operated.

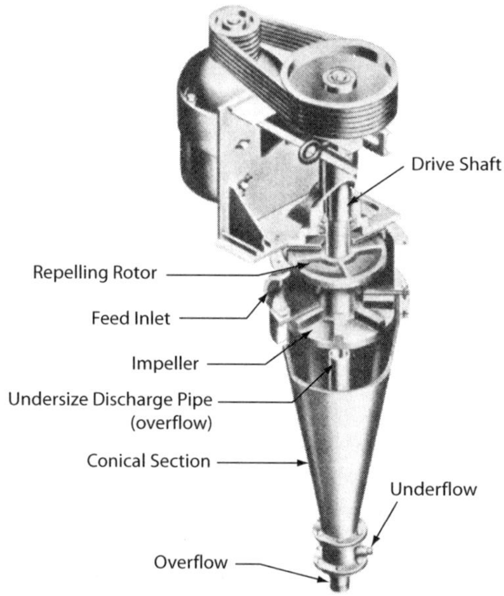

FIGURE 9.4 Centricone (Bradley 1965)

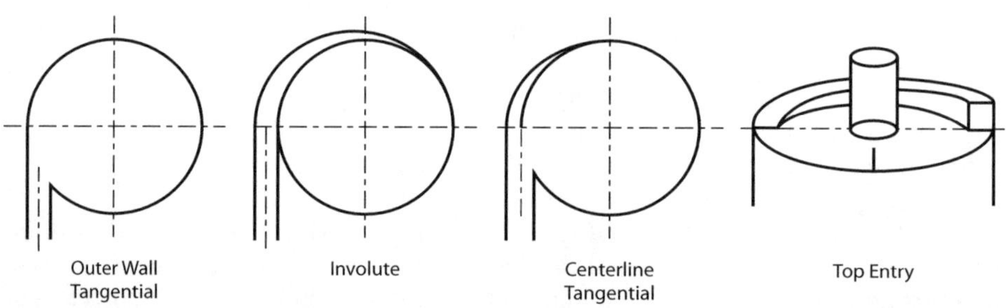

FIGURE 9.5 Feed entry designs for cyclones (Heiskanen 1993; reprinted by permission from Kluwer Academic Publishers)

In 1952, D.F. Kelsall published one of the first technical papers about how a cyclone operates (Kelsall 1952, 1953). This paper was the foundation for most of the future modeling work over the next 20 years. As the knowledge of cyclones increased, its uses and designs started to change. In 1953, R.C. Emmett and Dahlstrom published the first reference to open-top cyclones, which were later installed at Philex Mines in the Philippines and at an iron ore property in Minnesota. Also in 1953, H. Trawinski reported on the first use of cyclones in combination with centrifuges for starch separation (Trawinski 1955). He did a considerable amount of early work on cyclones in various industrial and chemical applications during the 1950s and was the inventor of the cyclones used by Amberger Kaolin Werke.

In January 1955, Equipment Engineers obtained a patent on the Krebs Model EE cyclone, which was a two-stage cyclone with the first-stage underflow feeding directly into the second-stage cyclone. This patent also included the involute-type feed for the first-stage cyclone. At this time also, the use of cyclones for closed-circuit grinding was starting to develop momentum in the United States—the Morenci concentrator being one of the first significant operations in the country in which cyclones were used in the grinding circuit (Barker, Papin, and Barr 1955).

Applications of cyclones outside closed-circuit grinding circuits continued to grow because they were small and cheap, and easy to install and operate. Here are some examples of applications that emerged in the period from 1955 to 1958:

- In the uranium industry they were used in the countercurrent washing process in which uranium in solution is separated from the tailings. Dorr Oliver had patented the use of multiple stages of cyclones for countercurrent washing.
- In the petroleum industry they were used for the control of drilling muds (Wuth and O'Shields 1955).
- In the clay industry small-diameter cyclones were used in South Africa to produce paper-coating clay and in combination with gravitational settlers to remove impurities from China clay (Hochscheid 1955; Naylor 1958).

Two important patents during this time were by Bradley (1957), who patented the use of two stages of cyclones to make liquid–liquid separation, and Sharples Centrifuges Ltd., which obtained a British patent (date unknown) that generally describes the advantages of using cyclones in combination with centrifuges and filters in chemical applications.

The potential for cyclones in closed-circuit grinding, however, was far greater than for other applications, and researchers continued to study cyclones for this process. The spiral or rake classifier was the standard classifier still being used at that time in most grinding circuits around the world, but as work progressed with the cyclone and early researchers installed test cyclones in grinding circuits, they found that cyclones had some significant advantages over classifiers:

- High-capacity and relatively small floor area for the larger tonnage mills.
- The percentage of solids in the coarse product that returned to the mill was close to the preferred grinding mill pulp density.
- Preferential classification of the higher specific-gravity middlings so that they were returned to the mill.

In copper concentrators, the cyclones provided a higher flotation feed density, and this helped to limit the volume of the flotation cells that were required for the larger concentrators.

For iron ore plants, cyclones were better able to provide the finer separations that were required to liberate the iron minerals for concentration and to produce suitable feed for the pellet plant.

Because cyclones preferentially placed the high specific-gravity middling particles into the coarse product for further grinding, the early work at copper concentrators in Chile and the United States showed that cyclones offered significant advantages in terms of copper recovery and feed tonnage over standard classifiers. As a result, copper concentrators were among the first to make full use of cyclones in grinding circuits. Some of the first installations of cyclones in copper grinding circuits occurred in 1955 at Acjoe in the Philippines, Majdenpek in Yugoslavia, and Chuquicamata in Chile.

The first major copper installation in North America was in 1956 at American Smelting and Refining Companies' Silver Bell Concentrator (about 40 miles north of Tucson, Arizona). Cyclones were tested in the grinding circuit in 1955–1956 when Russ Salter, mill superintendent, and Ed King, metallurgist, were open minded enough to replace relatively new twin 1.3-m spiral classifiers with four 0.5-m model D20B Krebs cyclones. Ironically, Dorr Oliver received the initial order for 0.6-m cyclones, but because of a very long delivery of 1 year, Silver Bell decided to cancel the order with Dorr and place it with Equipment Engineers. At the time the Silver Bell decision was made, the design of the

Pima Mining Company concentrator was being finalized. Pima was partially owned by Utah Construction (later Utah Mining), and its plant engineering group (now Fluor Daniel) was located in Palo Alto, California. Utah Construction's metallurgical department shared the test laboratory with Equipment Engineers, and they were kept well informed of the work at Pima. A last-minute decision was made by Utah Construction to retain the building design for classifiers but to install cyclones initially. If cyclones did not work, the classifiers could be put in. This led to the second major installation of cyclones in a copper concentrator and resulted in the second largest order of cyclones for Equipment Engineers.

The first major installation of cyclones in an iron ore grinding circuit was in 1955 at Reserve Mining (headquartered in Silver Bay, Minnesota; mine in Babbitt, Minnesota) where Dorr Oliver supplied Dorrclones that were 0.3 m in diameter. Early work on iron ore was carried out from 1951 to 1953 at pilot plants for Reserve Mining and Erie Company Mining (Auroa, Minnesota), and this proved that cyclones had advantages over spiral classifiers in the grinding circuit. In 1956, Erie Mining followed with an installation of cyclones with diameters of 0.375 m, which were built by the Gallagher Co. Following the early installations in copper and iron ore in 1955 and 1956, cyclones became the standard classifier in grinding circuits, and numerous installations followed in the next 10 years.

In 1957, Equipment Engineers split into two separate companies. Kelly and Dick Krebs retained the rights to the hydrocyclone and continued with Equipment Engineers. Clarkson retained the rights to the Clarkson reagent feeder and the slurry control valve and the Clarkson Co. became a very successful valve company. Equipment Engineers supplied most of the cyclones for grinding circuits in the late 1950s and early 1960s and in 1965 changed the name of the company to Krebs Engineers to honor Kelly and Dick Krebs.

Cyclones were being installed in numerous applications around the world in the late 1950s and early 1960s, and H.L. Bradley of the Sharples Centrifuge Co. investigated them in detail. His work resulted in many patents and, more importantly, in the book entitled *The Hydrocyclone*, which was published in 1965 (Bradley 1965). This book was the most comprehensive work ever published on the cyclone and today is still the bible for cyclones. Bradley was the first researcher to carry out detailed tests on many aspects of the cyclone including the effect of change in inlet, vortex finder and apex diameters, cylinder length, and cone angle. He also developed one of the first mathematical models of the cyclone, which other researchers later used to study the cyclone further.

Very Small Hydrocyclones

In the early 1950s, cyclones were being used in the kaolin industry to separate the finest kaolin particles into products that could be sold at a premium. Their success gave a boost to cyclone technology and led to the design of small cyclones that could efficiently extract the very fine particles from the feed. This was timely, because fine-grinding circuits were starting to become important.

Kaolin or China clay is a white clay whose main constituent is the aluminosilicate mineral kaolinite. Kaolin deposits result from the weathering and decomposition of granites and pegmatites. The size distributions of kaolinite and other clay minerals in the deposits range from about 30 μm to less than 1 μm (Figure 9.6). Other minerals in the deposits such as quartz, mica, and feldspar are coarser and have a minimum size of about 20 μm.

> *China clay is used in ceramic products, plastics, rubber, catalyst carriers, pigment additives, filler, and coating clays in paper. All these uses need different properties in particle size and form, brightness, gloss, reflectance, rheology, breakage*

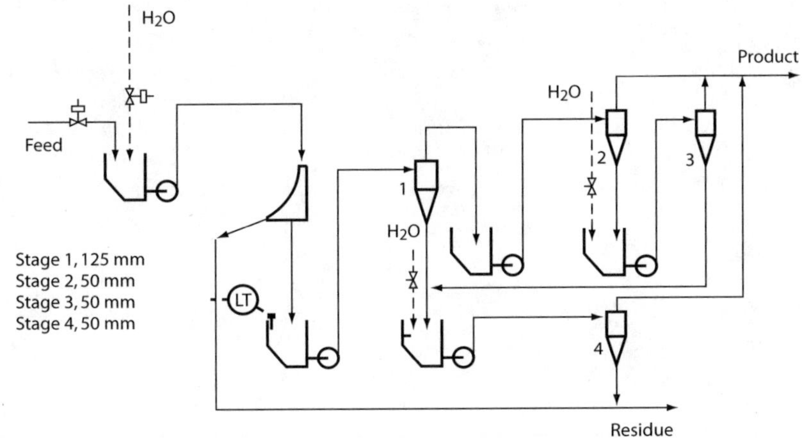

FIGURE 9.6 Flow sheet of a China clay production system using cyclones (Heiskanen 1993; reprinted by permission from Kluwer Academic Publishers)

> *strength and abrasiveness. Particle sizes vary from 90% −20 μm to 90% −1 μm. (Heiskanen 1993)*

Cyclones are the most suitable processing units, with 50- to 125-mm (2.0- to 4.9-in.) cyclones being used for the initial classification and 10- to 25-mm (0.39- to 0.98-in.) cyclones for the production of the highest grades of clay. Cyclones that can deliver −10-μm products are now important in processing circuits because of the very fine-grained ores, which must be ground and concentrated, and because of the demand for ultrafine particles of industrial minerals.

AIR CLASSIFIERS

The classification of very fine particles is carried out in many industries in air rather than water to avoid problems with chemical reactions of particles with water and with separation of ultrafine particles from water and drying them. Some of the disadvantages of dry classification are that it requires ducts and processing equipment that are capable of handling large volumes of air, so care must be taken to clean the air leaving the process to prevent particles from escaping into the atmosphere. Air classification has been used for thousands of years to remove waste from cereal grains and to concentrate fine particles of gold (see Figure 9.7 for an early application). It is now used in many industries to process thousands of millions of tons of solid particles annually. Cement and coal are the high-tonnage processes.

Air classification became important in the latter part of the 19th century when tube mills were developed to produce large volumes of portland cement, which had become a significant building material. The grinding process highlighted the need for efficient separation of ultrafine particles from the mill discharge, and screens proved to be inadequate. Mumford and Moodie patented the Askham separator in 1885 to classify very fine particles by air, and 4 years later Gebruder, which built cement manufacturing plants, improved the Askham separator and marketed it.

These machines had rotating parts for material and airflow distribution to improve the efficiency of separation. Its principle of operation is still used in modern air classifiers, or air separators as they are more commonly called. The Askham units were the first generation of dynamic air separators, and we will discuss their evolution over 100 years

FIGURE 9.7 (a) Miner in the 1930s using the wind to separate gold from ore after pulverizing (b) foot-operated air blower used for separation (Stone and MacKinnon 1976)

into second- and third-generation separators, during which time the rejection of coarse particles was improved and the bypass was minimized.

Another type of unit is the static air separator, which has no moving parts; the conveying air medium performs the separation by changes in air velocity and direction. The different types of separators will be discussed in this chapter. The change in separator designs had the objectives of better performance by

- Improving the precision of separation (eliminating misplaced particles)
- Reducing the amount of feed bypassing to the coarse product

This gives some control over the split size d_{50} and the width of the particle-size distribution, which is represented by its slope on a Rosin-Rammler-Bennett line (its sharpness). Table 9.1 shows the gradual improvement that has been achieved as air separators have been improved.

Static Separators

An early invention was a static air separator in which there are no moving parts and classification occurs by changes in air velocity and direction. Figure 9.8 shows the method of operation of a cone-type separator. The air stream (C) carrying the particles is converted from a directional flow through the outer cone into a rotating flow by guide vanes (E). The particles are subject to centrifugal force; the coarse particles move to the outer wall of the inner cone and spiral down it to enter the reject stream (B); and the fine particles move into an upward spiral in the center of the cone and are carried away (A, D). As air velocity increases, centrifugal force increases and separation becomes more efficient. The product size can be altered to some extent by changing the angle of the vanes, but the efficiency is low, and static classifiers can be regarded more as grit separators than efficient classifiers.

The V-Separator is a static separator that was patented in Germany in 1993 and in the United States in 1996 (Süssegger and Strasser 1996). The HPGR product required deagglomeration before separation into fine and coarse particles; the V-Separator, shown in Figure 9.9, does both duties.

In this configuration, the feed is directed down through baffles to meet the rising airflow. The weak compacts are broken as they collide with the baffles, and the fine and coarse particles are liberated. The fine particles are entrained in the air and swept

TABLE 9.1 Improvement in the performance of air separators for cement grinding circuits 1880–2000

Type		d_{50}, µm	Bypass, %	Sharpness
Static	0.25	Not applicable	60+	0.65–0.75
First generation	0.4	20+	50+	0.80–0.85
Second generation	0.55	15–20	15–35	0.85–0.90
Third generation	0.7	10–15	5–15	0.95–1.20

Source: Oner 2002.

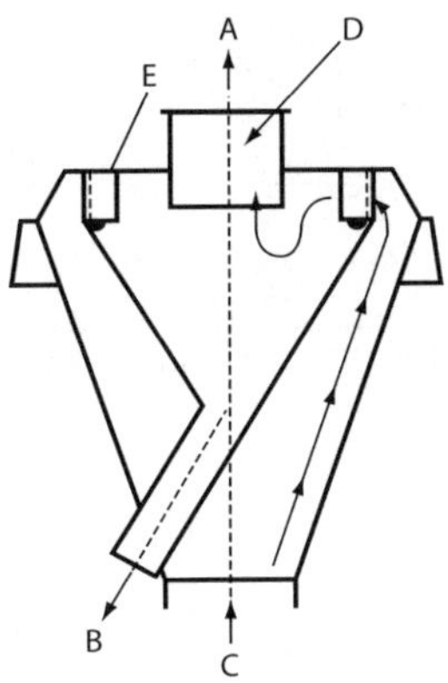

FIGURE 9.8 Static air classifier

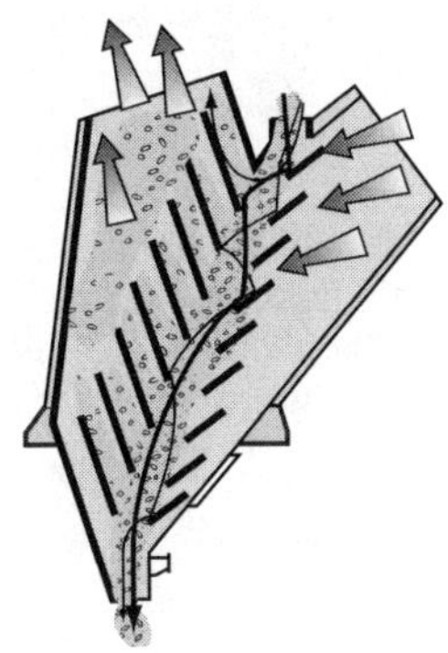

FIGURE 9.9 V-Separator (Lee and Süssegger 1995; reprinted by permission from KHD Wedag)

upward through a further set of baffles with which they collide, and the coarser particles are knocked out of the stream and fall back into the rising air. They either leave with the main stream of coarse particles at the base of the separator or enter the fine product. The separation size is adjusted by changing the airflow rate.

Like the static cone separator, the V-Separator is inexpensive, has low energy consumption, and is reasonably efficient at coarser sizes. Its place in a closed grinding circuit was described in the U.S. patent, and Figure 9.10 shows the sketch included in that patent. Separator feed is fresh feed and HPGR product; reject is HPGR feed; and fines are classified twice before a final product is made. The classifier rejects are ground in a ball mill. The V-Separator is representative of the type of unit in which separation occurs by changing the direction of flow frequently, thereby causing coarser particles to drop out of the stream.

Several separators based on this principle have been built, such as the Alpine Zig Zag classifier shown in Figure 9.11. These are known as grit separators.

Dynamic Separators

First Generation Dynamic air separators have several common characteristics:

- Internal rotors that are also feed distribution plates
- Airflows generated by internal or external fans
- Collision blades that interrupt the flow and rejection of coarse particles

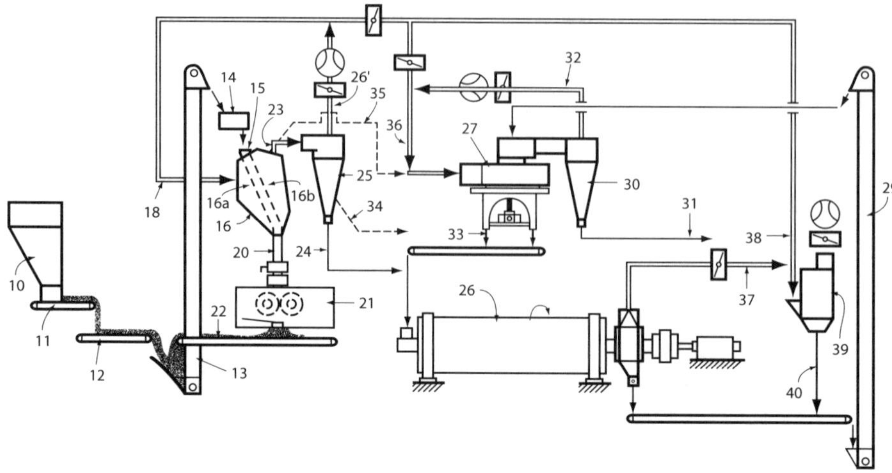

FIGURE 9.10 Sketch in the U.S. patent showing the V-Separator (item 16) in a closed grinding circuit (Süssegger and Strasser 1996)

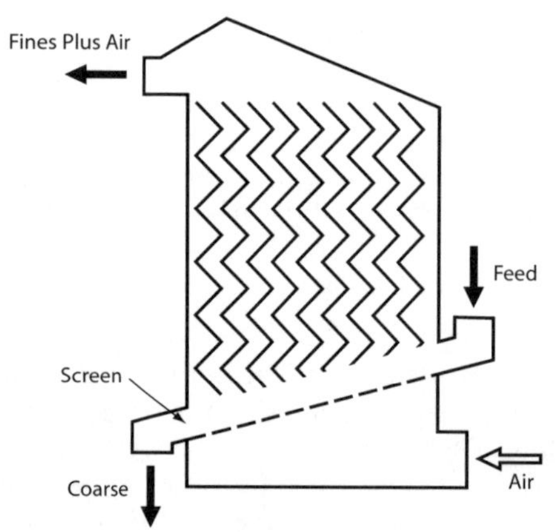

FIGURE 9.11 Alpine Zig Zag classifier showing the passages and collision surfaces (Klumpar, Currier, and Ring 1986; reprinted by permission from *Chemical Engineering*)

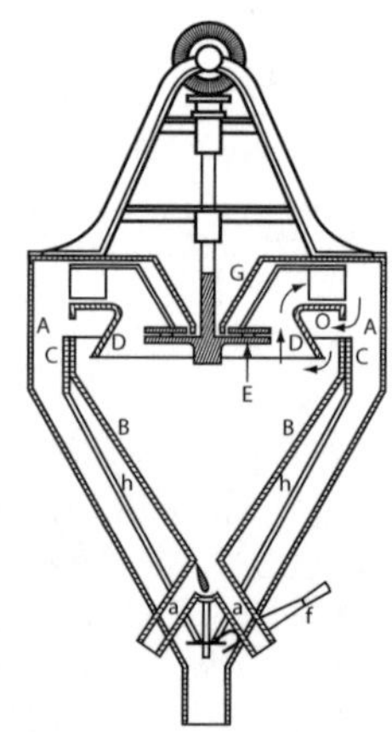

FIGURE 9.12 Askham air separator (Butler 1913)

Dynamic separators started with the Askham air separator, which became known as the first-generation air separators, and its principle of operation, which is the basis of modern air separators, is shown in Figure 9.12. D. Butler (1913) described the operation of this machine:

> *The machine consists of a funnel-shaped casing, within which is a second funnel with an annular space between the two. As its name suggests the separation is effected by a current of air which is produced by a fan of special design FF revolving in the upper or cylindrical part of the casing. The cement is fed into the cone G and falling onto the rotating disk E is thrown in a thin stream all round towards the fixed hood D. The current induced by the fan passes upwards and outwards between the fan blades carrying with it the finer particles which are thrown into*

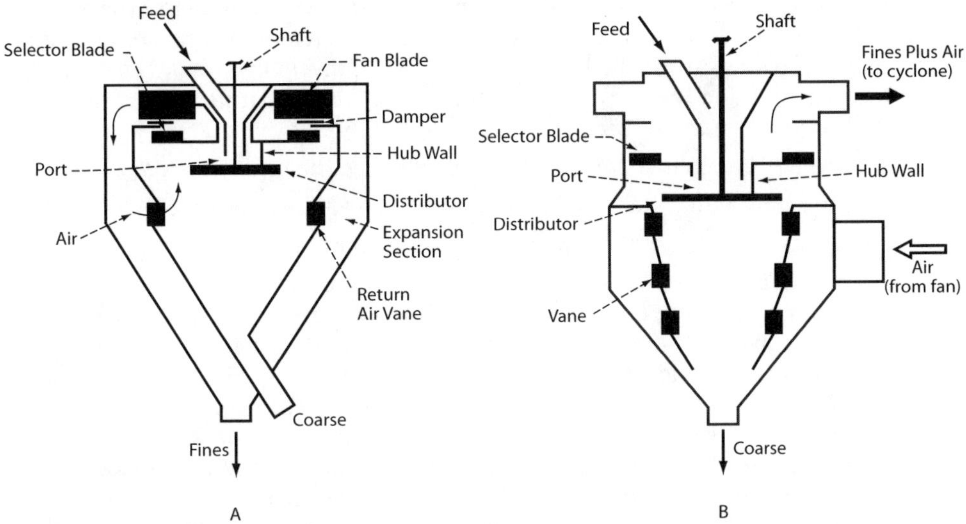

FIGURE 9.13 (a) First-generation Sturtevant Whirlwind (b) second-generation Humboldt-Wedag air classifier (Klumpar, Currier, and Ring 1986; reprinted by permission from *Chemical Engineering*)

> *the outer casing A and fall out the bottom for conveyance to the warehouse. The coarse particles which are too heavy to be lifted by the current of air fall into the casing B whence they return by the branch pipes aa to the mills for further reduction. The current of air returns through the opening o.*

The Sturtevant Whirlwind classifier, shown in Figure 9.13, was a popular first-generation classifier that was similar to the Askham classifier, although it had an additional auxiliary fan on the same shaft as the main fan. This created turbulence and assisted classification. Eventually the auxiliary fan had a separate variable-speed drive, which gave some control over the cut size.

The main advantages of the first-generation classifiers were good classification of very fine products on an industrial scale that had not previously been possible and relatively low capital costs. The main problems were

- Fine particles were not removed from the recycling air, and they accumulated in the coarse product, leading to a high recycle of finished product to the mill.
- The circulating air became very hot (>120°C), which is unsuitable for cement clinker circuits in which the gypsum could be affected. First-generation units have been preferred for circuits grinding raw materials for processing in the kiln.
- The sizing distribution and surface area of the fine product were difficult to change, although this problem was reduced when the auxiliary fan with the separate variable-speed drive was used.
- The distribution of feed on the disk and dispersion in the air were uneven, and this exacerbated the problem of high recycle.

For 75 years, this type of classifier was used in cement plants, and many are still in use today. The demand for more cement increased rapidly in the 1950s and there was a need to improve cement quality, which meant closer control over size distribution. The deficiencies of the first-generation air classifier had to be corrected, and this resulted in the design of the second generation of dynamic classifiers, which are discussed in the next section.

Second Generation First-generation separators suffered from high bypass, and it was difficult to change the sizing distribution and surface area of the fine product. In the 1950s, these problems outweighed the advantage of relatively low capital cost and had to be fixed. One solution was to remove the fines from the circulating air, and this led to a second-generation separator becoming available in 1960. Its main features were

- An external fan to circulate the air, which replaced the internal fan
- Several external planetary cyclones, which replaced the fines cone
- Independent control of the separator speed and air circulation

The air containing the fines went from the classifier to the cyclone where the solids were removed through the apex, and then to the fan that recycled it back to the classifier. The difference in operation of a first-generation Sturtevant Whirlwind air classifier and a second-generation Humboldt Wedag air classifier can be seen in Figure 9.13. The main advantages of the second-generation air classifier were better removal of the fines, sharper separation, lower bypass, and continuous control of the fineness. The main disadvantages were their large size, incomplete dispersion of the feed and removal of the fines from the recycling air, and a rather erratic cut size. For a few years after they came on the market, they were preferred to the first-generation units but they had a relatively short life because of their size and were replaced by the more compact third-generation unit.

Third Generation In the 1970s, a new dynamic, high-efficiency separator was designed with the objectives of decreasing bypass, giving a better precision of separation, and decreasing cut size (Onuma and Furukawa 1984). The O'Sepa separator was the first of the third generation, and it quickly proved its value when it was installed in a plant in 1979 (see Figure 9.14). Some improvements over previous models were

- The air entered the classifier horizontally and had a uniform velocity across the flow.
- The distribution plate was at the top of the airflow, and the feed fell as a well-dispersed curtain of particles.
- Particles passed through a rotating cage before entering the fine stream, and the collision of the particles with the bars of the cage assisted in rejecting the coarser particles from the flowing air.
- The product size could be adjusted online by changing the rotor speed.
- The dust-laden air from the mill was used as classifying air without impairing the classification efficiency, because the air left the separator and the fines were removed before it was recycled.

In operation the air passes through the stationary guide vanes, and the feed material is dispersed in the annular gap between these and the rotor. After passing through the vanes, the air moves in a horizontal vortex. The air carries the fine material tangentially across the face of the rotor that is turning in the same direction as the vortex. The coarse particles are separated by a combination of gravity, centrifugal, and impact forces and fall into the collecting cone at the base. The fine particles are conveyed to a dust collector. Their sharp classification and low bypass reduced the circulating load in tube mill–separator circuits and allowed an increase in feed rate by 20%–40%. The energy consumption per ton of cement was reduced by 15%–35%.

The principles involved in the O'Sepa third-generation or high-efficiency separator have been widely adopted by other manufacturers which have built particular features into their machines. The high-efficiency separators have replaced conventional separators

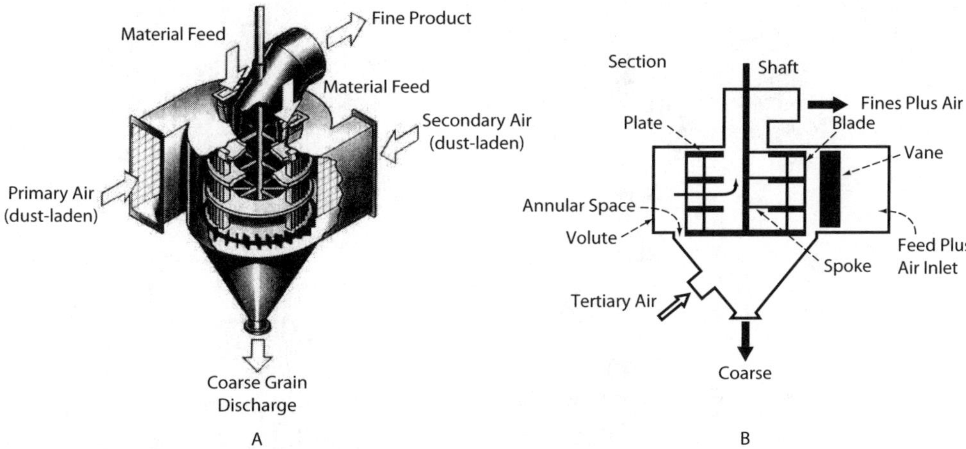

FIGURE 9.14 O'Sepa separator: (a) pictorial view (Fuller Company 2001; reprinted by permission from F.L. Smidth) (b) method of operation (Klumpar, Currier, and Ring 1986; reprinted by permission from *Chemical Engineering*)

in the housing of vertical roller mills and have reduced considerably the mass of rejects returning to the grinding table.

The Sepax separator is a high-efficiency separator which can deliver conventional fines and reject, along with very coarse or heavy particles (Figure 9.15).

The original Sepax separator (Figure 9.15a) was designed to work with air-swept ball mills with the feed that is the mill product already suspended in air when it enters the separator. Additional air enters the separator at its base, and classification occurs on the heavier particles that enter the bottom zone. The coarse reject, including steel from the mill, falls through the rising air and is removed at the base. Fine particles from the bottom zone and ball mill discharge are swept by air into the upper section of the unit where they are classified in a high-efficiency separator. The coarse particles from this separator are returned to the mill via an outlet above the bottom separator.

When HPGRs were used for grinding fresh clinker before it entered the tube mill, it became apparent that the grit separator in the base of the Sepax unit could be operated in closed circuit with the rolls to return coarse flakes to the rolls and send the fines to the high-efficiency separator for the production of finished cement. Figure 9.15b shows a Sepax unit with the base modified so that it contains the following:

- A feed inlet for the flakes discharged from the rolls and a deagglomerator rotor to break up the flakes
- An outlet for the grit (in this case the coarse product from the first separator) to be returned to the rolls
- A spiral inlet for the air

In all high-efficiency separators, the product can be controlled with the rotor speed or airflow although in practice the airflow is seldom changed.

Ultrafine particles have become essential in many industries, and it has been necessary to make classifiers that can separate at sizes of less than 10 μm, often much less. These cuts have been obtained by small, high-speed rotating cages that work in a similar manner to the rotating cages in the high-efficiency separators. This type of classifier will become more important as the demand for ever-finer particles increases.

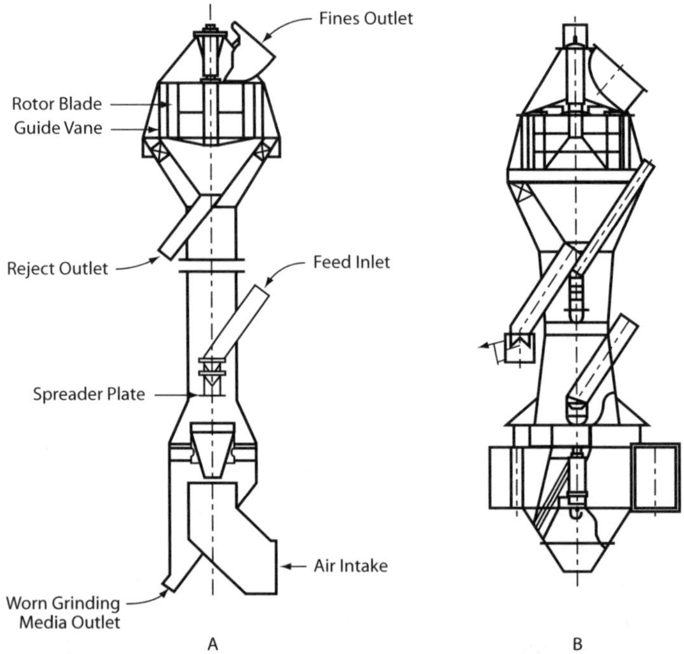

FIGURE 9.15 Sepax two-stage high-efficiency separator: (a) original unit with a single-feed entry, which is the tube mill product (b) modified for use with roller press product so that it has feed entries for the roller press and tube mill products (Smidth 2000)

CONCLUSION

Particle size separation using screens and classifiers became an integral part of size-reduction circuits. By separating the coarse from the fines in the discharge, the mechanisms came to give a measure of circuit performance, along with a means for controlling the circuit performance. The improved performance of the newest generation of air separators has added to the efficiency of grinding cement clinker. And the improvements to the integral air classifiers in roller mills has expanded their use for the dry grinding of cement raw material and other dry, nonabrasive minerals. Without the development of wet cyclone classifiers, the large-capacity wet-grinding circuits installed during the last 50 years of the 20th century would have required many more small ball-mill grinding circuits with either rake or screw classifiers.

To complete the history of size reduction, Chapter 10 discusses explosive methods for breaking rock from its natural formations to make feed for primary crushers. The breakage of rock is necessary to prepare the feed for plants that produce cement, aggregates, and metallic and nonmetallic ore products.

. . . .

CHAPTER 10

Explosive Rock Breakage

We have described how the processes involved in grinding developed in unison with new forms of energy and better materials: muscle power and stone, water power and iron, steam power and steel, and electricity and alloy steels. Explosive energy is the fifth form of energy, which is used to break rocks and to prepare them for further size reduction. The dangers and difficulties in blasting rock in mines have been great, and the priority of mining engineers has been to solve these problems rather than look downstream. It is only in the last 10 years that the development of precise drills, accurate explosives, and excellent sensors have given engineers the ability to understand and control explosive rock breakage, and they have used this knowledge to optimize blasting, crushing, and grinding to improve mineral extraction. Consequently, the objectives of blasts now include optimizing the feed to SAG mills, where the proportion of lump ore is important; minimizing the fines in products such as iron ores; enhancing the exposure of valuable minerals in deposits that are to be leached; and ensuring that the feed to primary crushers meets the required size. As knowledge increases, the link between blasting, crushing, and grinding will develop further.

The story of drilling and blasting, including the evolution of explosives from black powder to dynamite and ANFO, is told briefly in this chapter.

During the 18th and first part of the 19th century, the charging and firing of gunpowder shots was a somewhat risky procedure. A charge of gunpowder was poured into the shot hole. Sometimes a series of connected straws filled with powder or a train of powder laid in a paper twist were used to fill the needle hole (in the gunpowder). A piece of touch-paper that was supposed to burn for half a minute was lit and laid on the gunpowder trail, and the final performance of the shot-firer was to run as fast as possible (McAdam and Westwater 1958).

There are two reasons to break up large masses of rocks into fragments that can be readily transported away from the rock face:

1. To create space in the form of tunnels or surface excavations. For this purpose the broken fragments only have to be small enough to be transported from the rock mass for disposal.
2. To obtain broken rock for further processing. For this purpose the size distribution of the broken fragments is important because it affects downstream processing.

For thousands of years the sources of energy to break the rocks were fire, which cracked the surfaces, and human muscles, which wielded hammers and gads (see Chapter 3). Explosives were tested in the 17th century, and after initial problems blasting became the preferred technique to break rocks. Blasting was first used for military purposes in the mid-14th century, and there were good reasons that another 250 years elapsed before it was used for mining. Only a small amount of explosive was required to propel a missile, but moving masses of rock required very large amounts of explosives to be detonated. Because this was dangerous, miners were in no hurry to use blasting. But blasting was inevitable because the needs of mining and civil engineering left no alternative. The technology has now advanced to the stage at which the requirements of any civil or mining engineering application can be met accurately and safely.

The development of blasting technology has been difficult and dangerous:

- Blasting required progress in two areas in which the problems encountered were probably as tough to solve as any in the industrial world: drilling holes in rock and making all the components that are required in safe explosives.
- Explosives by their nature are dangerous, and innumerable tragedies occurred as progress was made in understanding and controlling blasting.

In this chapter it is possible only to briefly review the evolution of drilling and blasting, but this brevity does not reflect the importance of drilling and blasting in the spectrum of size-reduction processes. Indeed, it is the most complex and most difficult of all the breakage processes.

BLACK POWDER: THE FIRST EXPLOSIVE

The era of explosives started with the invention of fireworks some time in the first millennium AD. It is believed that fireworks started in China, one version being that they were created by the residents' habit of building wood fires on top of saltpeter rocks. The heat caused chemical decomposition of the saltpeter, and this generated copious quantities of gas and heat—a process called deflagration. The gas would have discharged particles from the burning saltpeter that glowed red hot, and this would have created a spectacular display. Addition of sulfur was found to improve the firing. These accidental fireworks led to the formulation of a saltpeter-carbon-sulfur mixture, which could be used for formal fireworks and for signals. Military use followed in the 8th century when a product made of 1 lb of sulfur, 2 lb of charcoal, and 6 lb of saltpeter was used for "...launching fire upon an enemy" (McAdam and Westwater 1958), but it was the 13th century before the great potential of controlled deflagration for launching missiles for military use was recognized and the 14th century before it was seriously used. Then explosives were developed for weapons and later for rock breakage, which changed the course of civilization.

The Rise of Black Powder

"Arabs are said to have invented the first real gun. In 1304 they reinforced a bamboo tube with iron and shot from it an arrow propelled by suddenly expanding gases released by ignited gunpowder" (Wilkinson 1966). The early formula for gunpowder was 10 parts saltpeter, 2 parts charcoal, and 1.5 parts sulfur (Eissler 1897)—a mixture that did not change much for 500 years. But it was not only the Arabs who were interested in gunpowder and guns; in Europe monks also worked on saltpeter as part of their efforts to build industries for their monasteries. Roger Bacon, an English Franciscan friar, published a formula for gunpowder in the 1240s which was similar to the Arab formula, and Berthold Schwartz in Germany, a monk who was probably in the Augustinian order, invented a gun about 1380 (Drinker 1888). Schwartz was "...credited with seeing the possibilities of developing the magical explosive as a propellant to hurl balls from cannons, the first of which he is said to have constructed" (Wilkinson 1966).

The military use of gunpowder had its critics. In the early 17th century, Francis Thybourel, the French poet, mathematician, and physicist, reportedly wrote an epitaph for Schwartz: "Here lies the remains of Berthold the Black, the most abominable of inhumans who by his invention has brought misery to the rest of humanity. Hence the term black powder was coined" (Atlas Powder Company 1987).

Military effectiveness has always been a powerful argument for developing better technology, and by 1520 black powder was being used in warfare to supplement traditional weapons such as the sword, ax, and battering ram. Mining skills were highly valued in armies at that time because of the importance of excavations during sieges, but even then mine owners were hesitant to use blasting for more than 100 years. Perhaps the reason was that explosives were still unsafe, a fact recognized by an edict issued by King James I in 1623 that prohibited the sale of defective powder and required all gunpowder to be approved and marked.

Despite these problems black powder was not to be denied. It was widely used in Hungary by 1642, and by 1687 it was being used in tin mines in Cornwall and at other mining and construction sites. At that time blast holes were 50–60 mm in diameter, 1 m deep, and took 1 kg of powder (Drinker 1888). C.G.W. Lock wrote a succinct comment in 1889 on the long-term effects of the introduction of blasting to mining:

> *The all-conquering Romans drove his slaves and prisoners by regiments to toil in the gold mines of Hungary, and miles of tunnels may yet be seen there carved in the solid porphyry by hammer and chisel, the mines being worked to this day. What the ancient Roman did by forced labour the Anglo-Saxon does now by the aid of steam, explosives and hydraulic power. (Lock 1889)*

Safety problems with using small amounts of defective powder in guns were bad enough, but the problems were much greater with the large amounts required to break a mass of rock. But the potential advantages for breaking rock were known to be great, and on February 8, 1627, blasting was first used in an underground mine. Caspar Weindl, a miner from Tyrol, gave a demonstration of rock blasting at the Royal Mines of Schemnitz at Ober-biberstollen in Hungary, which had a long tradition of mining dating back to Roman times. The blast was successful but "Powder for blasting did not become common for a long time thereafter for several reasons: boring implements were crude, mining methods were unsafe, and the cost of explosives was high" (Wilkinson 1966).

The value of black powder in engineering was that detonating it in a rock mass generated sufficient gas to heave boulders from the mass, provided the rock had existing fractures, although it was less useful for rocks without fractures. This enabled men to break boulders into pebbles at much higher rates than if the boulders were pried from the rock mass by hand. It tended to restore the balance between the rates at which rock could be mined and milled, which had been altered when water-powered stamp mills were introduced 100 years previously. In time it probably moved the bottleneck in the production of metallic minerals back to the mill. The black powder era for ore blasting lasted for 250 years until the rise of dynamite. During this time size-reduction processes for ores involved

- Black powder for the initial breakage of rock at the face
- Hand breaking the pebbles and boulders from the explosion
- Mechanical breakage of the finer fragments from blasting and hand breakage using stamp mills driven by water wheels

The value of black powder was such that it brought a seminal change to rock breakage technology and reduced greatly the terrible labor of manual mining. The tragedy of black powder was the cost of human life from the high rate of fatal accidents, and this cost had to be reduced.

Safety Fuses

The main cause of accidents was premature explosion. The chemical reaction that generated gas and heat from the decomposition of black powder was best activated by a spark or flame but getting close enough to the charge to apply a flame and escape quickly enough to avoid injury was a distinct skill, as noted earlier. Premature explosions could also occur accidentally.

In 1745, an electric spark was used by Dr. Watson of the Royal Society in England to explode black powder, and, in 1750, Benjamin Franklin improved his procedure by compressing the powder into a case (DuPont 1980). This could be done from a distance, which improved safety, and electric firing was tested successfully in Austria in 1804 and was being used in England by 1835 (McAdam and Westwater 1958). But electric firing was a step into the unknown that had to be taken cautiously with explosives. The main advance in igniting black powder is attributed to William Bickford in Cornwall in 1831 who placed a thin, slow-burning core of powder in a cable of jute and string and applied a flame at one end. The flame took some time to reach the explosive and this gave the shot-firer an opportunity to find safety. Within a few years the cable was coated with varnish to make it waterproof. The safety fuse reduced accidents by 90%. The fuse was continually improved by the use of better materials and these increased the safety factor still further. In 1958, fuses were burning reliably at a speed of 90 sec/m with an accuracy claimed to be 10%. It was far removed from the fire-and-run technique of 100 years previously.

EXPLOSIVES IN THE MID-19TH CENTURY

By this time, the Industrial Revolution was increasing the speed of social change; buildings and civil works were proceeding apace, and breaking and moving rocks was becoming a big business. Tunnels, roads, and canals had to be built through rocky terrains, quartz and limestone had to be mined to meet the needs for aggregate and metals, and stone blocks were required for buildings. As mentioned previously, the deflagrating explosive black powder was powerful enough to heave boulders away from rock masses with existing fractures but not powerful enough to shatter unfractured rock–a more powerful explosive was needed. Detonating explosives that decomposed rapidly and created very high gas pressures had been made in laboratories in Switzerland and Italy in the late 1840s, but testing stopped because they were so dangerous. The time for high-energy explosives had come but there were many problems to be solved.

Before continuing the discussion of high-energy, detonating explosives we will compare the characteristics of black powder and nitroglycerin, which was the first commercial detonating explosive. The rates of reaction of the two explosives were 400 m/sec for black powder and 1,700 to 9,000 m/sec for nitroglycerin. In fact the rate of reaction of nitroglycerin is so fast that if a pipe 8.5 km long is filled with nitroglycerin and detonated by a blasting cap, the entire column would be converted into gas within one second (Peele 1941). Both explosives generated huge volumes of gas that were increased many times by the heat of reactions, which were up to 5,000°C. The result of these differences in practice was that black powder heaved and moved the rock; nitroglycerin also shattered it.

Nitroglycerin was calculated by Alfred Nobel to be more than 13 times as explosive as black powder when both heat and gas were taken into account (Eissler 1897).

NITROGLYCERIN: THE FIRST HIGH-ENERGY EXPLOSIVE

Nitroglycerin, an oily fluid produced by reacting glycerin with nitric or sulfuric acid, was discovered by Ascanio Sobrero, a professor at the University of Turin in Italy, in 1847. Sobrero found "...its properties so unpredictable and frightening that he warned against its use" (DuPont 1980). Other discoveries that were important in formulating high-energy explosives are described in the following paragraphs.

Mercury fulminate was made in 1799 by reacting an excess of alcohol with a solution of mercury in nitric acid (Karrer 1947). The salt exploded violently when struck or ignited and was soon used as a detonator in percussion caps (DuPont 1980). This was used in percussion caps in guns in 1814, but it was another 50 years before it was used in rock blasts.

Guncotton was the product of reactions between nitric and sulfuric acids and cotton. It was made by Christian Schonbein, a professor from the University of Basle in Switzerland, in 1845 and was more powerful, safer, and less corrosive than black powder (DuPont 1980). Its main use came when it was dissolved in nitroglycerin and formed a jelly that was used in gelatin dynamites.

By 1850, sufficient research had been done for a high-energy, detonating explosive to be made, but nothing happened until Immanuel and Alfred Nobel (see Chapter 2) recognized the potential value of nitroglycerin in 1859. Their interest in high-energy explosives caused them to investigate their manufacture and use, and they were sufficiently impressed by nitroglycerin that they borrowed 100,000 francs to build a factory in Sweden. It was easy enough to make the oil but there were many problems in using it. For example, the oil had to be poured into fissures in the rocks (mechanical drills were unavailable), and there were losses of explosive in the rock masses and danger in the undetonated remnants of the broken products. In addition, potentially dangerous leakage occurred during storage and transport.

Alfred Nobel systematically investigated the possibilities for making the best use of nitroglycerin. After experimenting with detonating mixtures of nitroglycerin and black powder under various conditions, he concluded that the full effect of nitroglycerin as an explosive could only be obtained by detonating it with a compound such as mercury fulminate that was exploded in close proximity (Drinker 1888). The market for nitroglycerin explosives was brisk but premature explosions still made safety a serious problem. In 1867, Nobel observed that the mineral kieselguhr packed around containers of nitroglycerin absorbed leaking explosive. He made a solid nitroglycerin-kieselguhr mixture containing 25% kieselguhr (McAdam and Westwater 1958) into cartridges which themselves had strong explosive properties. He called the product dynamite and demonstrated that dynamite could be handled safely and detonated when required. With its discovery "The blasting of roads and mines now entered a new era, as did the military art" (Halasz 1959). In another development Nobel combined guncotton with nitroglycerin to make a very powerful explosive that he called blasting gelatin.

Dynamite was quickly found in practice to be effective for breaking and moving rocks because of its high detonation velocity and explosive power. The two explosives, black powder and dynamite, were extensively used until 1920 when dynamite started to take over the market and became dominant by 1950.

Initiation of Explosions

Reliable initiation of the explosive decomposition of nitroglycerin was a problem from the start. Black powder safety fuses were tested but gave erratic results, so Nobel returned to a detonator that he had patented in 1867 before dynamite was invented. This was a tin capsule containing mercury fulminate that exploded when struck, and when used to detonate dynamite it markedly improved its reliability. During the next few years Nobel produced several strengths of dynamite by using different amounts of nitroglycerin and nitrocotton, but even with the blasting caps there were still problems with accurate and reliable detonation. So inventors turned to electrical detonators.

Electric detonation had been used intermittently since its invention by Watson in 1745; wires contacted the explosive in the cap made incandescent by an electrical current, thereby starting the detonation. Early electrical detonators acted instantly and it was not until 1895 that the delay detonator was invented by H. Julius Smith, who put a short piece of fuse between the ignition charge in the cap and the base charge, thereby creating a delay. This gave miners the ability to detonate charges in a series rather than instantaneously, and eventually sequential explosions were found to be more efficient for breaking a mass of rock than a single, large explosion.

But caution was always the rule when changing blasting practice, and it was not until 1910 that delay detonators were accepted as safe for routine use. Many improvements have been made since then: Detonators are available with any delays that are required; and the scatter in firing times has been reduced to a negligible level.

The Dynamite Era

The dynamite era lasted for 100 years, and during this time the volume of explosives used to break rock increased greatly. Copper ore is a good example. In 1900, 500,000 tons of copper were produced from ore containing about 2% copper, which means that about 25 million tons of ore were mined. In 1913, the Utah Copper Company alone milled 25,000 tpd of copper ore (about 10 million tons per year). This was blasted in open pits, and much more waste was also blasted to give access to the ore. Production rates at open-pit mines grew ever larger as ore grades declined and demand for copper increased, which meant more drilling and blasting. Dynamite broke the rocks efficiently, but the dangers of unexpected explosions and the problems of headaches and nausea that occurred when nitroglycerin came in contact with unprotected skin were always present. It was time for another explosive to be developed that was strong enough to break large volumes of rocks and was cheap, safe, and hygienic. Ammonium nitrate proved to be the answer, and when problems concerning its solubility in water were solved, its sales increased and the dynamite era declined quickly.

Data showing the rise and decline of the dynamite era and the rise of the ammonium nitrate era are given in Table 10.1.

ANFO: POWERFUL AND SAFE

As Table 10.1 shows, there was a spectacular rise in consumption of ammonium nitrate explosives during the 1960s. This was the start of the ammonium nitrate–fuel oil (ANFO) era in industrial explosives, which is expected to continue well into the future.

Ammonium nitrate was made by J.R. Glauber in 1659 (DuPont 1980). Its explosive properties must have been known early because of its connection with sodium and potassium nitrates, but it was 200 years before they were recognized commercially. In 1867, Ohlsson and Norrbin in Sweden patented an ammonium nitrate–nitroglycerin mixture that when sensitized with fuels formed an explosive that was less powerful but cheaper than dynamite. Alfred Nobel purchased the patent and expanded his range of products by using ammonium nitrate in dynamites. It was soon used in other mixtures; for example, in France it was mixed with di-nitro-naphthalene, and cylinders of the mixture were coated with paraffin, which contained carbon and repelled water. Because it was a safe, inexpensive, and strong explosive, sales increased. The problem was that it was soluble in the water that often flowed into drill holes and destroyed its sensitivity. This problem had to be solved if ammonium nitrate explosive was to be used in wet environments. The simplest solution, which was patented by DuPont in 1934, proved to be the best: Package the ammonium nitrate fuel mixture with the detonator in sealed cans. After that, this explosive became widely used in quarries and open pits.

Ammonium nitrate was regarded as a safe explosive because it was difficult to detonate, so the tragic events of 1947 came as a terrible shock and showed what a devastating explosive it really was. In that year shiploads of bagged fertilizer-grade ammonium nitrate (FGAN) blew up, causing great destruction and loss of life in three locations: Texas City in the United States, Brest in France, and the Black Sea near Russia. These terrible

TABLE 10.1 Sales of explosives in the United States, 1915–1975, in thousands of pounds

Year	High Explosives	Ammonium Nitrate Explosives	Black Powder
1916	289,840	—	215,575
1921	212,086	—	160,021
1926	378,203	—	157,687
1931	257,735	—	79,830
1936	309,906	—	81,698
1941	422,469	—	59,458
1946	499,491	—	36,824
1951	719,494	—	13,985
1956	1,003,458	—	5,593
1961	533,663	696,202	1,521
1966	613,495	1,486,104	463
1971	316,373	2,237,524	117

Source: Adapted from DuPont 1980.

accidents highlighted both the danger and the great potential of ammonium nitrate when used as a commercial explosive.

FGAN was cheap so mining engineers started to experiment with it to break rocks. Results were inconclusive, probably because the coarse crystals and grains that were available were harder to detonate than powder and the carbonaceous sensitizers may not have been mixed well. Tests continued and within a few years FGAN prills (2 mm in diameter) were being made that quickly absorbed the fuel oil used as a sensitizer. This led to producing ANFO explosives, which soon replaced dynamite in dry holes because it was safer, cheaper, and easier to handle. The speed of hole charging increased tenfold, because ANFO could be blown up into holes and compacted. ANFO needed a big kick for detonation and this was provided by a small plug of dynamite that was detonated using a blasting cap.

By the early 1960s, the advantage of ANFO in underground as well as surface blasting was clear:

> *Up holes were previously limited to 70 ft (20 m) because of the problem of hand tamping conventional explosives, but pneumatic charging of ANFO overcame this restriction and resulted in improved blasting, reduced handling and improved charging performance. (Davies 1967)*

The problem of wet holes increased as mines grew deeper. The revolutionary step in solving this problem was to make use of the water rather than try to repel it. This was done by making a saturated solution of ammonium nitrate in water, dispersing more ammonium nitrate and sensitizer in the solution, and initiating the dispersion with a powerful booster. The concept of an ammonium nitrate-water-sensitizer mixture was created by Melvin Cook of the University of Utah and out of it came the new technology of water-based explosives—measurably improving mining blasting and size reduction.

Cook, discussed in Chapter 2, was a brilliant physical chemist who worked on explosives with DuPont and the U.S. government before and during World War II. We rank his work with ANFO alongside the work of the cyanidation and flotation engineers as the great technical achievements in the mining industry in the 20th century.

Cook's early research at the University of Utah was about the theory of mineral flotation, but in 1952 he returned to explosives research after being appointed director of a project supported by the U.S. army. He understood well the potential and problems of ammonium nitrate because he worked on it with DuPont.

> *Cook's greatest commercial explosives invention was formulated in December of 1956 while consulting for Iron Ore Company of Canada where he created a new blasting agent using an unusual mixture of ammonium nitrate, aluminium powder and water. (Khodorovskiy 2000)*

This was a seminal discovery in mining and civil engineering, because the blasting agent was the first slurry explosive and it allowed the safe ammonium nitrate to be used conveniently in wet ground. This explosive consisted of an ammonium nitrate/water dispersion sensitized by aluminum powder that produced a large amount of heat when oxidized. Guar gum was used as a thickener to minimize segregation and leakage. Water-based explosives were immediately successful because of their low cost, high strength, safety, and convenience. Problems using them in large-diameter up holes were eventually solved by adding cross-linking agents that instantly increased their viscosities.

A great advantage of water-based explosives was that they could be produced in bulk and pumped, which reduced the charging time and more accurately prepared mixtures.

> *For example at one mine 140 tonnes were loaded in 11 shifts and at another mine 13 tonnes were loaded in one shift. This compares to typical ANFO loading rates of just 2 tonnes per shift. (Law et al. 2001).*

The result was that by 1963 twice as much ammonium nitrate was being used for blasting in the United States as high explosives, and this increased to 10 times by 1975. An important factor in making stronger, cheaper, and safer explosives was the development of high-energy emulsions. These consisted of droplets of ammonium nitrate in water covered by oil and distributed in finely dispersed voids in gas bubbles that acted as hot spots to initiate the chemical reaction. Emulsions could be blended with ANFO on site to produce explosives with different detonation properties. A report by engineers at one mine noted that

> *In 1993 an up-hole emulsion based explosive using gas microballoon sensitization was introduced. This product was specifically formulated for use in up holes and trials showed that it could be retained in a 127 mm vertical up hole. ...the product was used routinely in up holes with diameters from 76 mm to 115 mm. (Law et al. 2001)*

The change from dynamite to ANFO had advantages as far as health, safety, and the rate of hole charging were concerned but it was not without problems. When it was first used it was mixed on site with fuel oil, and the mixing was often less than thorough with adverse effects on blasting. Because of its solubility in water, many novel methods were devised in those days to keep water away from ammonium nitrate—one of the more interesting examples being at one open pit where rain fell for 2 days out of 3 and molasses was added to ANFO to repel water. This practice was not continued; perhaps the cure caused more trouble than the problem. Despite the teething problems, ANFO-based explosives were safe and efficient by 1990, and any required detonation properties could be obtained. The safety hazards, nausea, and headaches associated with nitroglycerin had been eliminated.

But better explosives by themselves were not enough to ensure that best results were obtained by blasting. It was also necessary to optimize drilling and initiation systems.

Initiation Methods

An initiation system starts the detonation of the main charge, after which firings should occur at the exact times that are planned to ensure best results. As discussed earlier, the objective of initiation systems during the black powder and dynamite eras was to improve the safety of explosives. By 1900, the first electric delay detonator was available. This detonator was a fusehead that started the burning of a black powder fuse connected to an ordinary detonator. The delay time before detonation was about 8 sec, so it improved safety but it also reduced the scatter in firing times of nominally identical initiation systems from about 8,000 to 600 μsec. During the 20th century, progressively larger blasts required the sequential detonation of many charges, and research on initiation systems turned toward minimizing the scatter in firing times. The timeline for the main developments was

- 1920s: Instantaneous electric detonators worked by initiating a tiny amount of a very sensitive primary explosive such as lead azide that, when detonated, initiated a small amount of a sensitive secondary explosive in the main charge. Placing the primary explosive in the main charge would have jeopardized safety but was needed to detonate the secondary explosive. The scatter was reduced to 100 μsec.
- 1950s: Short-delay detonators allowed blasts at intervals of 25 m-sec. Over 10 years the scatter was reduced to about 6 m-sec.
- 1970s: High-precision blasting caps improved the control of vibration, fragmentation, and damage.
- 1990s: Electronic detonators can deliver a defined delay time and can be made with a scatter of less than 1 m-sec.

The driving forces to improve initiation systems were safety and the need for more rock to be broken by explosives. In the 1970s, multiple ring blasts or mass blasts were introduced to underground mines and they greatly enhanced productivity by permitting long, uninterrupted campaigns of drilling and blasting. The detonation of huge amounts of explosives in sequence in a single mass blast, however, meant that any initiation problem could cause serious production losses. Significant improvements were required to electronic initiation systems to ensure correct timing, avoid misfires, and maintain an acceptably low level of ground vibration. Consequently, the 1970s and 1980s were periods of intense research into electronic initiation and vibration monitoring, and blast design and associated fragmentation. Parallel R&D programs investigated the application of electronic, computing, and detection systems, and ultra-high-speed photography supported this research. The result was that, by 2000, initiation systems were available that were accurate and controllable.

DRILLING

Rock Drills in the 19th Century

Humans have been drilling holes in rocks for more than 4,000 years. Drill holes used for explosives need to penetrate some distance into the rock mass; otherwise all that happens when the explosive is detonated is that rock at the surface is shattered. A steam-driven rotary drill was built in Cornwall about 1815, and a steam-driven churn drill, based on the ancient Chinese method of lifting and dropping a rod tipped with a bit, was built by Isaac Singer in the United States about 1835 to work on the Illinois-Michigan canal. Neither drill became popular, and Singer turned his attention to a mechanical sewing machine for which he achieved more notoriety.

Manual drilling with hammers and bits proved to be the only reliable method of drilling until the invention of the mechanical percussion drill in the mid-19th century. Three men could drill a 2-m deep hole in granite in 5–6 hours on average depending on the conditions (Chugh 1985). In 1850, manual drilling with hammers and bits was still the only reliable method of making holes in rocks (see Chapter 3).

The first mechanical percussion drill was invented by J.J. Couch in Philadelphia in 1849 (Rodengen 1995). The principle was that steam was admitted alternately to each end of a cylinder; the drill was thrown like a lance at the rock on the forward stroke, caught and drawn back on the reverse stroke, and thrown again (McAdam and Westwater 1958). It was the first drill that did not depend on gravity, but it was cumbersome and the machine weighed several thousand pounds. Another drill patented in 1851 by Couch's former colleague J.W. Fowle (Rodengen 1995) was an improvement because the drilling rod was attached to the piston and impelled by the entire power of the engine. This drill, which was also powered by steam, was used successfully in practice, and a similar drill built the same year in France used either steam or compressed air. In these percussive drills the bit made contact with the rock every cycle of the engine and this was how drills worked for 40 years. By 1855, several events gave impetus to the development of mechanical drills.

Work had started on the 8.2-km Hoosac Tunnel in Massachusetts (Sumberg 1999) and in 1857 on the 14.5-km Mont Cenis Tunnel in the Italian Alps (*Harper's New Monthly Magazine* 1871). Manual drilling was known to be very slow and expensive; initial manual drilling advanced 23 cm per day at each end. It was estimated that with manual drilling the Mont Cenis Tunnel would take 30 years to complete, so there was much interest in pneumatic drilling.

Compressed air had become recognized as a suitable method for transmitting power over long distances. It had been used with furnaces to melt metals since 2500 BC, and the first mechanical compressor was built in 1776, but it was the need to transmit power over long distances to drive rock drills that made compressed air an important industrial tool.

The number of mines in operation and civic works under construction was increasing and the need for drilling was growing rapidly.

The growing interest in drilling led to many innovations and inventions such as

- Simon Ingersoll's invention in 1871 of a steam-driven drill mounted on a tripod that kept the drill steady and enabled it to be operated at virtually any angle (Rodengen 1995).
- Germain Sommeiller's use of waterpower from rivers near the Mont Cenis Tunnel to drive air compressors to power the pneumatic drills. He proved the value of compressed air in rock drilling, and the mechanical drills he used greatly reduced the time to build the tunnel. This success came despite his experience that for every nine rock drills that were in operation on each heading, 54 machines were being repaired. Eventually he was able to transmit power by compressed air over 7 km and improve the rate of drilling with 280-kg drills from 0.46 m per day in 1861 to 2.27 m per day in 1870.

A major advance in drilling came in the late 1880s when C.H. Shaw, an engineer in Denver, Colorado, devised a hammer drill in which the piston was separated from the drill rod and hammered it each cycle, thereby increasing the frequency of blows and the rate of penetration. This was suitable for up holes when the cuttings fell out by gravity but not for down holes, and this problem was not solved until 1897 when J.G. Leyner patented a hollow drill rod through which air was pumped and flushed out the cuttings. Water soon replaced air, which reduced the dust that had until then been a lethal problem.

Success in building tunnels using pneumatic drilling and blasting by dynamite had a great impact on rock engineering and size reduction, although the work was still extremely hard. In 1878, percussion drills using steam or compressed air could drill holes 87–100 mm in diameter and up to 16 m long, but the weight of the drills, about

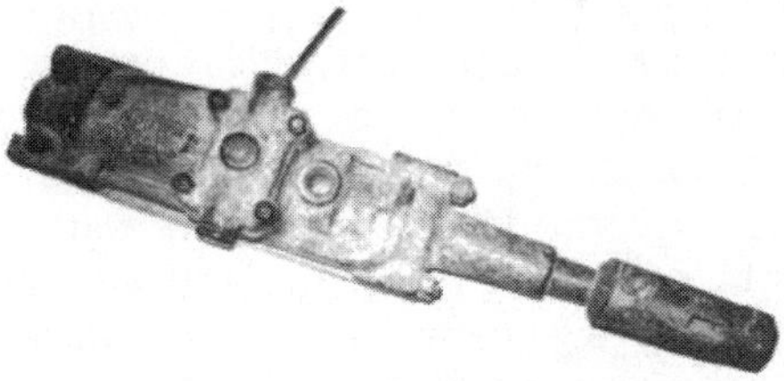

FIGURE 10.1 A rock drill used in Maldon, Victoria, in 1890 (Drinker 1888)

FIGURE 10.2 Light rock drills driven by compressed air in the 19th century (Foster 1894)

180 kg, meant that they were hard to transport and set up (Drinker 1888). Figure 10.1 shows an 1890s drill.

Lighter pneumatic drills were built that could be handled by one man, and they increased the rates at which minerals could be mined in confined spaces. The drills may have been lighter but were still difficult to handle, as shown in the sketch in Figure 10.2.

Drilling and blasting advanced quickly during the last quarter of the 19th century because extensive R&D programs were driven by the need to reduce the costs of the great tunnels and canals being built in Europe and the United States.

Some of the problems with building them were

- Drills had to work in an environment that was as difficult for a machine as could be imagined, and it was necessary to build machines that were stronger, more durable, and more reliable. It has been said that no tougher service for steel is known than in a rock drill. During every minute of operation there are approximately 2,000 violent collisions between the piston and the shank end of the drill rod.
- Explosives were still dangerous and unpredictable, and their safety had to be improved without reducing their breakage energy.

The R&D programs carried out from 1875 to 1900 were difficult and dangerous, but with all their problems they were very successful and resulted in explosive rock breakage becoming one of the great technical achievements of the 19th century.

Advances in Rock Drilling in the 20th Century

By 1900, rock drilling had reached the point at which hammer drills were available that operated at several hundred cycles per minute, worked at any angle of inclination, had the cutting head held at the rock face by an air cylinder, and had the chips and fine detritus continually removed from the working zone by air or water.

Compressed air was being used to power the drills because it was an inexpensive and efficient method of transferring energy over long distances or to difficult sites, such as underground. But there were some problems:

- Machines were heavy and had to be fixed in place with the main support being wedged between the floor and the roof. Two men were required for each machine.
- The dust created by drilling was a health hazard. This could be reduced but not eliminated by the water used to flush away the cuttings.
- The cutting heads that were part of the drill rods lasted only about 0.5–1.0 m in hard rock and up to 3 m in softer rock before requiring sharpening.

Keeping the drills sharp meant there was a constant flow of drill rods between the mining face, the cage in which equipment was transported up and down the shaft, and the drill shop that was always built on the surface close to the shaft to reduce the travel time. At one gold mine in Canada at which 1,500 tph of very hard rock were mined, 34,000 drill steels were sharpened each month (Fisher 1937), and many men were employed to work on sharpening and transporting drill rods. In some mines the forged cutting end of the drill rod was replaced by a detachable bit, but this was not always a success because of breakage problems at the connection point, which was inherently weak. The bit could also wind off the rod and be lost.

Much effort went into building light, hand-held drills that had become available by 1912. These had a profound effect on drilling because they made the work of the driller underground much easier and eventually eliminated the heavy tripod mountings that were so difficult to transport and erect.

Before the 1950s, drilling underground was usually performed with one-piece drill steels and relatively small percussion drills. These steels were commonly used in batches of increasing length and decreasing bit diameter to drill holes of 6 m or less. An early improvement was the development of tungsten carbide bit inserts that greatly reduced the rate of bit blunting. Tungsten carbide was made in Germany in the late 1920s, and cemented carbide bits, in which pieces of tungsten carbide were embedded in steel, were made in Sweden in the early 1940s, mainly for use in civil engineering projects. They were introduced into the mining industry in the 1950s. Detachable tungsten carbide bits and coupled rods permitted the drilling of 20 m long holes because the size of the drill drive no longer limited the length of the drill rod. Larger percussion drills were produced with new rod feeding and handling systems, and over this development period blast-hole diameters increased from 30 to 75 mm.

An inherent shortcoming of the coupled rod percussion drill system was the flexibility and discontinuity of the drill string. This led to bending of the blast hole, which increased exponentially with depth and caused erratic and unpredictable spacing of the hole toes, unevenness in charging density, and poor fragmentation. Consequently, blast hole length was limited to 25–30 m. Another shortcoming was the loss of hammer energy between the out-of-hole drill and the bit, again due to drill string flexibility and looseness of the drill string in the hole and of coupling threads.

In the 1970s, in-the-hole hammer drills and rotary drills were introduced to large underground mines. With in-the-hole hammers, percussion is delivered directly to the bit and less energy is wasted in the drill string (Figure 10.3); with rotary drills, large-diameter,

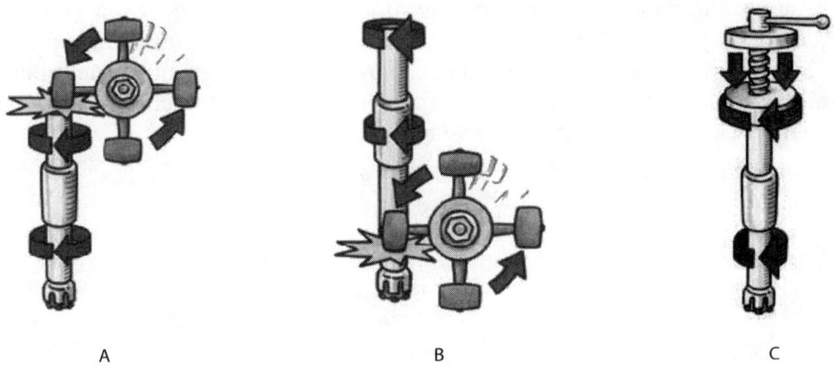

FIGURE 10.3 Methods of operation: (a) out-of-hole hammer drills (b) in-the-hole hammer drills (c) rotary drills (Ingersoll-Rand Company 1995; reprinted by permission of Ingersoll-Rand)

smoothly coupled drill strings are used that fit neatly into the blast holes and hole deviation is greatly decreased.

The result of these developments was that drill holes increased from 57 mm in diameter and 1–2 m long in the 1960s to 150 mm in diameter and 30 m long in the 1980s. The main objective in introducing these large blast holes was to permit use of longer holes and to increase sub-level spacing (a system of horizontal underground workings from which ore is mined). To achieve uniform ring charging density the spacing between blast hole toes had to be widened accordingly.

Unlike in open cuts where bench drilling permits the use of parallel blast holes, underground limitations usually require the use of ring patterns, which are far from ideal for even distribution of explosives. Drill holes in ring patterns underground also require collars to be uncharged, thereby wasting portions of drill holes. By the late 1980s, sub-level spacing of 50–60 m became possible, which allowed larger blasts and more efficient blasting, but even in the 1980s drill holes deviated considerably from plan. This meant that not all the explosive was placed according to the blast design.

In the 1990s, advances in mechanical and materials engineering, electronics, and computer systems made drilling machines more durable and more reliable in terms of drilling to plan. Straighter drilling has been achieved by the use of tube drilling and guide tubes, more rigid drilling rigs, more accurate (computer-controlled) initial alignment of the drilling machine, and down-hole hammers. TV cameras and survey instruments provide details of where the holes went. Drill holes still deviate but their accuracy is good enough to be able to investigate and understand the effect of drill-hole variables on blast results.

BLASTING: A PART OF TOTAL ROCK BREAKAGE SYSTEMS

Advances in drilling and blasting were always hard won, and the work was onerous and dangerous until well into the 20th century. Objectives of blasting during the 20th century were

- Before 1960: to make excavation possible
- 1960–2000: to make excavation efficient
- After 2000: as part of the overall rock breakage process

Much of the technical progress during the 1960s and 1970s was aimed at increasing productivity by gaining economy of scale. This did not always benefit fragmentation.

Larger loading and haulage equipment were introduced, and extraction developments and draw-points were designed with larger cross sections. Pneumatic-tired machines powered by diesel engines were introduced, providing greater flexibility and permitting the handling of larger rock fragments. From a mining consideration alone the main requirements were that rock fragments were small enough to

- Avoid choking draw-points or bridging brows
- Be handled readily by loaders, trucks, and tipples
- Fit into crushers

Better equipment made larger rocks easier to handle, and this improved mine production rates, but there was still a need for good fragmentation because this improved the ease of flow of the ore and the maintenance of equipment. The effective cost of poor fragmentation can be several times the cost of the blast itself because of problems such as

- Excessive dilution of the ore
- Increased secondary blasting
- Reduced loading rates
- Difficulties in ore handling and transport
- Poor performance in the mill comminution system
- Variable ore quality in concentration (Scott, Chitombo, and Kleine 1993)

In economic terms, if the cost of a blast is $1.30/ton, the extra cost caused by poor fragmentation can be $1.50/ton in grinding and $2.00/ton in concentration.

Modeling Blasting Processes

Blasting is a complex process and many variables are involved. Consequently, a model of blasting that predicts fragmentation accurately is necessary to optimize total size-reduction systems. The first theory of blasting was proposed in 1792 and a model based on it was used to design blasts (Drinker 1888), but its objectives, details, and how well it worked were not known. Bond was the first engineer in the 20th century to write a model to predict the sizing of broken products in open cuts, and several engineering models of fragmentation have been developed since then. The more important, listed by developer and data required, follow:

- Bond (1959): mean block size, energy input
- Favreau (1970): detonic and physical rock properties, blast design variables
- Kuznetsov (1973): powder factor, rock mass classification, explosive parameters
- Harries (1977): blast vibrations, dynamic rock properties
- Dinis Da Gama (1983): rock structure mapping, energy input, comminution behavior
- Cunningham (1987): rock mass parameters, mean block size, blast design details
- Kleine (1988): in-situ block size distribution, energy distribution, breakage characteristics (Scott, Chitombo, and Kleine 1993)

These models were valuable aids in improving blasting practice when they were used by engineers who were familiar with them. The problem was that procedures for determining the composition, fracture pattern, and breakage characteristics of the rock mass being blasted were not well developed, and without this knowledge the interactions between the rock mass and the explosive could not be understood or controlled accurately.

Since the 1990s, there has been an increasing move toward the application of numerical techniques to model the complete blasting process that includes detonation, breakage, fragmentation, and displacement (Preece, Jensen, and Chung 2001; Dare-Bryan, Wade, and Randall 2001; JKMRC 2001). Continued improvements in computing technology will make the future application of these types of models feasible. In addition to the development of these models, the advances being made in other technologies, such as measurement while drilling, blast monitoring systems, and image analyses, are providing the necessary data to validate the new generation of blasting models. The next 10 years should see a significant improvement in our ability to model the blasting process. This progress will enable blasting to fit into the spectrum of size-reduction processes that extends from blasting through crushing and grinding processes to ultrafine grinding.

CONCLUSION

This history of the development of explosives that can be used to break rocks shows the potential of explosives to become a useful means for controlled size reduction in preparing feed for primary crushers and mineral extraction processes. Currently, the mining industry needs methods to best utilize the energy made available from blasting, which requires that methods be developed for describing the structure of the rock formation to be mined. This becomes part of the future for more efficient size reduction of ores.

CHAPTER 11

Size Reduction in the 21st Century

The art of prophecy is very difficult, especially with respect to the future.
–Attributed to Mark Twain

NEW SCENARIOS FOR SIZE REDUCTION

For more than 10,000 years, size reduction has been the unending effort to produce ever-larger volumes of flour for consumption, rocks and pebbles for building, and fine mineral particles for producing cement and metals. The main barriers to creating large-scale size-reduction systems, tools, and machinery were the amount of energy that a machine could utilize and the quality of those materials used to build the tools and machinery.

In the latter part of the 19th century, during the Industrial Revolution, inventions were abundant and steam was found to be superior to both waterpower and muscle power. The availability of energy and the quality of available materials to build machinery and tools, however, still limited the growth of size-reduction technology.

During the first half of the 20th century–a time of world wars and economic depression–interests other than size reduction were predominant. It was not until later that the excitement of innovation again gripped the industrial community. Engineers then began to use all the tools at their disposal–electricity, new materials, and computers–to improve all size-reduction processes and machinery.

Now, as the 21st century begins to unfold, the technical environment in which size-reduction processes operate has changed in a number of important ways.

- Energy is available in unlimited amounts. If it is not available at a particular location, power systems can be extended to the site or a generator can be obtained.
- Materials now available for constructing the tools and machinery have excellent resistance to wear and impact and are continually being improved.
- Motor size and the availability of gears and pinions no longer are a barrier to construction, enabling the creation of size-reduction machines with capacities unimaginable even 50 years ago.
- Explosions now can be designed to shatter millions of tons of rock.
- Machines have been developed to meet numerous special requirements.

The demand for size reduction will continue to grow. The question is: What course will the technology take? Making the assumption that the path of progress followed during the last 100 years will continue unchanged probably is not prudent because of the equipment and process limitations–whether anticipated or unrecognized–that have developed.

The history of size reduction shows that, as the demand for size-reduction systems, tools, and machinery has developed, the available technology using available energy sources has met the demand, and often also has been accompanied by *new* processes and machinery. At the same time, the need for human muscle power has been reduced. The expansion created by the Industrial Revolution, however, also increased contamination of the air, water, and soil to the extent of creating potentially harmful effects. The solutions to these environmental and societal problems also could create new demands on size-reduction technologies, systems, tools, and machinery.

FUTURE PROSPECTS

The prospects for the future of size reduction cannot be precisely established. The history and the questions facing those industries that use the technology, however, indicate the areas requiring study. Technology has always stepped up to meet the demand for size reduction. Educational and research facilities have been created to teach and to search for a better understanding of current processes and to develop new, more efficient ones. Although the size-reduction industry faced financial, environmental, and social stumbling blocks in the latter half of the 20th century, the need for continued study never has been more apparent.

Energy

In these early years of the 21st century, there appear to be few limits on the fuels necessary to drive electric generators. The environmental, political, and financial problems associated with the fossil and hydrocarbon fuels used to generate the steam that drives the turbines needed to drive the generators, however, could limit the use of these fuels in the years to come. Hydroturbines seem to be a clean solution—until the magnitude of the construction of dams and large lakes necessary to obtain the necessary water levels is considered. Additionally, the use of nuclear energy is, of course, subject to many environmental and public safety objections.

The availability of the needed energy, then, could be problematic, and reduction and rationing of the fuels or the electricity may prove to be necessary. If this does happen, more and more efficient size-reduction processes will be needed to make use of the available energy.

The production of fines with the current size-reduction processes is inefficient in the use of energy but is economically efficient in processing large continuous flows of heterogeneous materials in circuits that are both easy to operate and easy to control. The efficient continuous flow of material will be required for any new commercial reduction process developed.

Scale-Up

The scale-up to larger equipment can create problems—particularly when the degree of increase for capacity and machine volume are not the same. An example of this is the scale-up of wet-grinding overflow ball mills. The power drawn by ball mills operating at the same: (1) percent of volumetric loading; (2) percent of critical speed; and (3) length of grinding compartment varies as the ratio of the mill diameter is raised to the 2.3 power $(D1/D2)^{2.3}$. This formula applies when the power drawn by one mill diameter is known. It keeps the three key variables in the mill power equation at the same level with only the mill diameter and power as variables. The mill volume varies as the diameter squared.

As the mill diameter increases, the volume of the mill available per unit volume of the feed decreases. As ball mill diameter changes for a constant percent of critical speed, mill speed in revolutions per minute changes at inverse ratio of the square root of the mill diameter. The volume of the media in the mill is a function of the ratio of mill diameters squared. As the mill diameter increases, both the mill speed in revolutions per minute and the number of balls per unit of feed decrease, and the number of ore and media contacts decreases. As ball mills become larger in diameter, the largest mills become more inefficient—shown either by a decrease in feed rate needed to produce the specified grind or by a coarser grind at the designed feed rate. These scale-up factors are the same in all tumbling mills—mill power or mill capacity is directly proportional to mill length.

As the diameter of ball mills increase, the retention time of the feed in the mill decreases and the flow rate of the slurry through the mill increases. At this time, when large ball mills become more inefficient, the remedy is to reduce the volume of the ball charge, which increases the mill volume available for the feed. It also decreases the power drawn by the mill, which reduces the capacity of the mill. For each ore and fineness of grind, a method is needed to determine both a minimum retention time required to produce the desired grind and a maximum flow rate of the slurry (that will not disturb the action of the grinding media). Currently, this information can be learned only in the size of the mill to be put in the concentrator. Because of the scale-up factors, minimum retention time and maximum slurry flow rate cannot be determined from small ball mills in a pilot plant.

Simulation, the newest calculation method, uses data from a smaller mill to simulate the performance of a different sized mill. For accuracy, simulation for tumbling mills should take into account the effects of the scale-up factors.

Technology

Each new generation of computers—particularly personal computers—increases the availability of and the capabilities for creating simulations for all sizes and types of size-reduction processes and machinery. Computers will continue to be important tools in studying size reduction and in making the needed developments.

Manufacturers of mechanical machines and electrical motors have demonstrated the capabilities and readiness necessary to engineer and build larger equipment for size reduction. The scale-up factors can be projected, as they affect the size-reduction performance of the machinery more than the mechanical performance of the machine. The technologies are available to design and manufacture new processes and machines as they are developed.

CONCLUSION

The prospects for the future of size reduction vary from the continuation of the currently available processes, circuits, and machinery to a complete revision and new processes and machinery. The breakage methods of compression, impact, abrasion, and shearing seem to be the practical mechanisms available; however they all require energy and are subject to wear on the working surfaces of the machinery. Additionally, energy of some type always will be required for size reduction.

A developing approach is the in-situ leaching of gold ores and oxide minerals. Leaching of ore could be done in place, crushed in stockpiles, and overburden could be put in stockpiles, and tailings put in special leaching ponds. The chemicals used in leaching and the liquid-solid separation create problems, however, particularly with the contamination of groundwater and free-flowing streams and rivers in the area of the leaching basins.

The direction of future prospects is dependent upon demands created by environmental and economic pressures, the character of domestic and international politics, and the availability and type of educational and research facilities. The only certainty for the future of this field is the existence of a continuing need for size reduction to develop and fit the lifestyles of people both today and in the future.

Glossary

Aggregates – hard, inert materials used for mixing in various-sized fragments with a cementing material to form concrete or mortar.

Air separators – a machine where air is the carrying medium instead of water, and particles are separated by centrifugal force. Static separators are similar in operation to hydrocyclones, but dynamic separators have internal rotating mechanisms that can be controlled to ensure that the separation is as sharp as possible and that a minimum amount of material passes through the separator unclassified.

ANFO – ammonium nitrate–fuel oil. The most widely used explosive for mining in the late 20th century, because it is inexpensive, safe, powerful, and easy to handle, it will continue to dominate explosive rock breakage operations for many decades.

Autogenous mills – ginding media for rocks of various sizes and generated by shattering and abrasion of the ore being ground.

Ball mills – machines that use separate grinding media—steel balls, cast iron balls, or, sometimes, ceramic balls. The ball charge normally contains a mixture of sizes related to both the size of the feed and the required product size.

Black powder – an explosive used for fireworks for many centuries. The first explosive used for breaking rocks in mines, this deflagrating explosive is made from potassium nitrate, composed of saltpeter, carbon, and sulfur. Its speed of reaction is 400 m/sec.

Bradford breaker – a tumbling mill to break coarse coal that is brittle and shatters easily when dropped. The coal enters the feed end, is lifted by the rotating mill, and shatters by impact when it falls. The mill has holes in the cylindrical section through which small particles of the shattered coal leave the mill. Larger particles are repeatedly dropped until they shatter while the tougher shale does not shatter and leaves the mill at the discharge end.

Classifiers – machines in which broken particles are separated into fine and coarse (or light and heavy) streams.

Colloid mills – high-speed mills in which rotating pins pass each other with little clearance between the pins, creating intense turbulence and shear forces in the slurry close to the pins. See also *pin mills*.

Compartment mills – ball mills divided into compartments by vertical diaphragms. The first compartment has large balls to grind coarse particles; the second one has finer balls to grind smaller particles. The apertures in the division heads retain balls and larger particles in the compartment. Occasionally, a mill has three compartments. Used mainly as dry mills for grinding cement clinker, they are also used for grinding refractory gold ores prior to roasting. The grinding media in cement clinker mills are usually alloy steel balls. In wet-grinding compartment mills, alloyed steel rods are occasionally used as the grinding media in the first compartment.

Cone crusher – high-speed gyratory crusher with the crushing chamber designed to accept a finer feed than a conventional gyratory crusher and produce a finer discharge.

Crushers – machines that reduce the size of boulders and pebbles by compression and impact.

Deflagrating explosives – low-energy explosives that burn rapidly with intense heat. Large volumes of gas are generated by the chemical reactions.

Detonating explosives – high-energy explosives that decompose so rapidly that they explode. The copious volumes of gases produced by the chemical reaction have very high pressures and are efficient and dangerous explosives.

Disk mills – machines where stirring is carried out by disks attached to a shaft rotating at high speed and used for very fine grinding.

Dynamite – a solid explosive that is much safer than nitroglycerin by itself, which is more difficult to confine and handle. Dynamite contains 75% nitroglycerin and 25% kieselguhr.

Edge mills – machines that have a stone or concrete disk rotated on a stone base. Cereal grains are ground between the moving rim of the disk and the fixed stone base. The edge mill was the first type of roller mill. Horse-powered edge mills are still used in many small villages where they are used to grind small amounts of grain each day.

Explosives – a mixture of chemicals that produces heat and a large volume of gas when they react. The gas exerts rapid pressure on walls that confine it. Explosive rock breakage occurs when the explosive is placed in rock and detonated and the gas moves or shatters the confining rock.

Fire setting – building fires against rock walls and keeping them burning for hours until the rock heats enough to crack so that it can be removed easily.

Grain roller mills – counter-rotating rolls used in grain mills. The break rolls cut most of the bran from the endosperm, the scratch rolls break composite bran-endosperm particles, and the reducing rolls grind purified semolina to finished flour. The roll surfaces may be corrugated or smooth depending on their use.

Gyratory crushers – machines that are similar in principle to jaw crushers but the fixed plate is the inner surface of an inverted cone. The moving plate is a cone that fits into the fixed cone. The moving surface is mounted on a large shaft called the main shaft, which is suspended from the top. The surfaces of the cones converge. The center cone rotates and gyrates, as its surface moves toward and away from the fixed cone during each revolution. During the gyration, rocks are caught in the wedge between the plates, and crushed and broken fragments are released.

Hammer and gad – a heavy stone used to drive a lighter, wedge-shaped stone into a crack in a rock to detach a boulder from the rock wall.

Hammer mill – high-speed rotating hammers that break rocks by impact. Broken particles are flung against a screen, the fine particles pass through the screen and are discharged, and the coarse particles are retained for further breakage. In an out-of-hole hammer drill, or drifter drill, each blow is struck at the end of the drill string out of the hole, and energy is transmitted through the drill rods and couplings to the bit. In a down-the-hole hammer drill, the hammer piston is at the bottom of the hole and directly contacts the bit.

Hand stones – moving stones that almost touch fixed stones and are moved by muscle power. Cereal grains and ore particles are ground between them.

High-pressure grinding rolls (HPGRs) – Spring-loaded counter-rotating rolls that do not touch but which break pebbles and particles by exerting immense pressure on a bed of particles. The product emerges as a sheet of agglomerated particles that must be deagglomerated prior to further processing.

Hydrocyclones – cylinders into which slurries are introduced tangentially at high speed and rotate around the inner wall. The coarser or heavier particles are held by centrifugal force at the wall and spiral toward one end of the cylinder where they are diverted by a cone to an apex from which they are collected. The finer or lighter particles form a spiral in the center of the cyclone, which rotates in the same direction but moves in the opposite direction, and are collected through a vortex finder. Centrifugal force is the separating mechanism.

Jaw crushers – machines that have two converging steel plates, one of which is fixed and the other which moves a short distance back and forth. Rocks fall into the wedge between the plates as the moving plate recedes, then are crushed when it moves forward, and broken fragments are released when it recedes again.

Jet mills – machines with fixed cylinders into which very high speed jets of particles are introduced so that they collide and cause the particles to break autogenously by impact and abrasion.

Mortar and pestle – the size-reduction device that has remained unchanged in principle since the Stone Age. The pestle (rod with a flat-surfaced hammer) is moved up and down in the mortar (bowl-shaped vessel), striking the base of the mortar during each cycle and breaking particles caught between the mortar and pestle.

Nitroglycerin – the first high-energy explosive. An oily liquid made by reacting glycerin with nitric or sulfuric acid, it is a detonating explosive, with speed of reaction up to 9,000 m/sec.

Pebble mills – machines for grinding media for pebbles from stream beds or natural deposits or for pebbles formed by breakage of coarse rocks.

Pin mills – high-speed machines in which rotating pins pass each other with little clearance between the pins, creating intense turbulence and shear forces in the slurry close to the pins. See also *colloid mills*.

Portland cement – made by burning a mixture of lime and clay.

Querns – a moving stone traveling in a rotary motion. Solid particles are added through a hole in the center of the moving stone and are broken as they travel to the periphery; the ground product is then discharged from the periphery.

Rake or spiral classifiers – tanks into which ore-water slurries are fed, with the coarser particles settling and being removed by a rake or spiral and the fine particles overflowing the top of the tank and being removed in a launder. Gravitational force is the separating mechanism.

Rittinger classifier – the original hydraulic classifier. The slurry flowed horizontally across settling boxes of increasing size, with the coarsest particles settling in the first box and finer products in later boxes.

Rod mills – machines that have grinding media for alloyed steel rods that usually are the same length of the mill shell.

Roll crushers – spring-loaded, counter-rotating rolls in which rock pebbles are broken by compression individually and no bed of agglomerated particles is formed.

Roller mills – machines that have steel cylinders attached to horizontal rotating shafts. There are two types of mills: (1) those in which particles are caught between counter-rotating rolls and broken as they pass between the rolls (used to grind cereal grains and small rock pebbles); and (2) those in which particles are caught between a rotating roll and the bowl with which it is in contact. This type of mill usually has an air separator built into the structure with the rollers. The separator classifies the product from the rollers and returns the reject for further grinding. It is used to grind soft rocks such as coal, limestone, and phosphate rock.

Rotary rock drills – rotating a drill bit against the rock, penetration is caused by applying pressure to the top of the bit. Soft rock requires lower feed pressure and faster rotation speed than hard rock.

Saddlestones – horizontal fixed stones, over which a moving stone travels in a back-and-forth motion.

SAG mills (semiautogenous mills) – machines where steel balls occupy up to 12% of the mill volume as supplemental grinding media. The steel balls are needed to break critical-size particles—those that are too small to be broken effectively by abrasion and too large to be broken effectively by impact. The name SAG mill is currently used for all primary mills whether fully autogenous or semiautogenous.

Shorthead cone crushers – machines designed to produce a very fine discharge.

Size reduction – the breaking of materials such as grains, pebbles, rocks, and ore to a smaller size using available tools or machines driven by available sources of energy.

Stamp mills – mechanized mortars and pestles driven by waterpower from about 1500–1850, then by steam power. The main machines used for fine crushing or coarse grinding of ores for 400 years were replaced in the 20th century by fine crushers and ball mills.

Stirred mills – machines where the material being ground is kept in motion by stirring rather than tumbling. Grinding media and particles acquire kinetic energy by stirring with an impellor or by agitation. Particles are broken by abrasion and impact when they are caught between grinding media, colliding with themselves or with the wall. Stirred mills are used to make very fine particles. The general rule is that higher-speed mills use smaller grinding media and produce finer particles. Vertical high-speed stirred ball mills are called peg mills, and horizontal high-speed stirred ball mills are called disc mills.

Tower mills – vertical cylindrical machines in which the stirring is carried out by screw-type impellors operating at relatively low speeds.

Tumbling mills – machines with horizontal rotating cylinders containing grinding media and particles to be broken. The mass moves up the wall of the cylinder as it rotates and falls back into the toe of the mill when the force of gravity exceeds friction and centrifugal forces. Particles are broken in the toe of the mill when caught in the collisions between the grinding media themselves and the grinding media and the mill wall. Mills may be batch or continuous. In continuous mills, feed enters one end and broken product leaves the other. In batch mills, the material to be ground is charged to the mill, which is then closed and run until the material is ground. Then the mill is stopped and the load in the mill is dumped out of the mill and another load is put in. Usually, mills are identified by the type of grinding media. In tumbling mills, the grinding media and particles acquire potential energy that becomes kinetic energy as the mass falls from the rotating shell.

Vertical roller mills – today's version of the age-old edge mill. They use up to four very large rollers which rotate in a spinning bowl. Feed enters the center of the bowl, is ground as it flows between the rollers and the bowl, and is discharged from the periphery of the bowl. It is normally entrained in an upward-flowing air stream and classified in a separator. The reject returns to the center of the bowl where it is ground again.

Vertical-shaft impact crushers – machines with a feeder rotating at high speed inside a fixed cylinder. The feeder is designed to accept rocks at its center and propel them from its outer rim at high speed so that they are flung against the fixed cylinder and broken by impact. The wall of the fixed cylinder is designed to hold compacted broken rock to minimize wear.

Vibrating and nutating mills – machines with very small balls kept in rapid motion in a confined volume by moving the mill shell at very high speed.

Water mills – mechanized querns used for grinding cereal grains. They were replaced by roller mills in large-production flour mills starting about 1890 but are still used in many villages located on rivers and streams and in small mills that produce stone-ground flour.

Work index – a parameter developed by Fred Bond that defines the resistance of a material to size reduction in a machine. The kilowatt-hour per ton required to reduce the material from theoretically infinite feed size to 80% passing 100 μm is calculated from plant data but can also be predicted from laboratory tests that measure the grindability of the material.

References

Agricola, G. 1950. *De Re Metallica.* Translated by H.C. and L.H. Hoover. New York: Dover Publications.

Atlas Powder Company. 1987. *Explosives and Rock Blasting.* Dallas: Atlas Powder Company, a subsidiary of the Tyler Corporation.

Australian Wheat Board. 1998. *The Australian Grain Industry.* Melbourne: Australian Wheat Board.

Barker, L.M., J.F. Papin, and R.C. Barr. 1955. *Application of the Liquid–Solid Cyclone as a Classifier in Closed Circuit Grinding at the Morenci Concentrator.* Washington, DC: American Mining Congress.

Bartley, B. 1996. Inventing our crusher. Speech to the Institute of Quarrying New Zealand Inc., February, in Auckland.

Barton, D.B. 1961. *A History of Copper Mining in Cornwall and Devon.* Cornwall, UK: Truro Bookshop.

Bass, E.J. 1988. Wheat flour milling. In *Wheat: Chemistry and Technology.* Volume 2. Edited by Y. Pomeranz. Saint Paul, MN: American Association of Cereal Chemists.

Beedell, S. 1975. *Windmills.* New York: Charles Scribner's Sons.

Benedict, C.H. 1952. *Red Metal: The Calumet and Hecla Story.* Ann Arbor, MI: University of Michigan Press.

——. 1955. *Lake Superior Milling Practice.* L'Anse, MI: Michigan College of Mining and Technology.

Bennett, R., and J. Elton. 1898–1904. *Corn Milling.* Volume 2. London: Simpkin, Marshall.

Bickle, W.H. 1958. *Crushing and Grinding: A Bibliography.* London: Her Majesty's Stationery Office.

Biringuccio, V. 1990. *The Pirotechnia of Vannoccio Biringuccio: The Classic Sixteenth-Century Treatise on Metals and Metallurgy.* New York: Dover Publications.

Bond, F.C. 1952. The third theory of comminution. *Transactions SME/AIME* 193:484–494.

——. 1959. The work index in blasting. *Quarterly Colorado School Mines* 54(3):77–82.

——. 1961. Crushing and grinding calculations. *British Chemical Engineering* 6(6):378–385.

——. 1964. An expert reviews the design and evolution of early autogenous grinding systems. *Engineering and Mining Journal* August:105–111.

Bradley, D. 1957. British Patent 8,740,17. December 4.

——. 1965. *The Hydrocyclone.* London: Pergamon Press.

Bretney, E. 1891. U.S. Patent 453,105. May 26.

Broadbent, S.R., and T.G. Callcott. 1956. A matrix representation of breakage. *Journal of the Institute of Fuel* 29:528–539.

Brown, R.L. 1941. Generalized law of size distribution. *Journal of the Institute of Fuel* 14:129–140.

Brundiek, H. 1989. The roller grinding mill—its history and current situation. *Aufbereitungs-Technik* 30(10):610.

Butler, D.B. 1913. *Portland Cement—Its Manufacture, Testing and Use.* London: E and FN Spon.

Calvin, W.H. 1991. The throwing madonna: Essays on the brain. http://www.williamcalvin.com/bk2/bk2.htm. Accessed February 2003.

Casson, L. 1966. *Ancient Egypt*. Amsterdam: Time-Life International (Nederland) BV.

Chugh, C.P. 1985. *Manual of Drilling Technology*. Rotterdam: A.A. Balkema.

Clutton-Brook, J. 1952. *Horsepower–A History of the Horse and the Donkey in Human Societies*. Cambridge, MA: Harvard University Press.

Coghill, W.H. 1929. Fine grinding in the tri-state. *American Zinc Institute* 12(5/6):100–117.

Coghill, W.H., and F.D. DeVaney. 1938. *Conclusions from Experiments in Grinding*. Rolla, MO: Missouri School of Mines and Metallurgy.

Cook, M.A. 1974. *The Science of Industrial Explosives*. Salt Lake City, UT: Ireco Chemicals.

Crushing and Grinding. 1931. Genesis of the Hardinge mill. *Crushing and Grinding* 1:47.

Cunningham, C.V.B. 1987. Fragmentation estimates and the Kuz-Ram model: 4 years on. *Proceedings of the 2nd Symposium on Rock Fragmentation by Blasting*. Keystone, CO: Society of Experimental Mechanics.

Dahlstrom, D.A. 1952. High efficiency desliming by use of hydraulic water additions to the liquid–solid cyclone. *Transactions of AIME* 193(August):789–793.

Dare-Bryan, P., L. Wade, and M. Randall. 2001. Computer modeling of bench blasting for grade control. *Proceedings of the 27th Annual Conference of the International Society of Explosives Engineers* 2:13–24. Cleveland, OH: International Society of Explosives Engineers.

Davies, E. 1967. Mining practice at Mount Isa Mines, Australia. Pages A14–A40 in *Transactions of the Institution of Mining and Metallurgy*. London: Institution of Mining and Metallurgy.

de Camp, L.S. 1961. *Man and Power*. New York: Golden Press.

——. 1970. *The Ancient Engineer*. New York: Ballantine Books.

Del Mar, A. 1917. *Tube Milling*. New York: McGraw-Hill.

Demir, O. 1998. *Cappadocia: Cradle of History*. Ankara, Turkey: Ajans-Turk Publishing Company.

Dennehy, M.J., and S.K. de Kok. 1953. Application of liquid–solid cyclone as classifier in closed circuit grinding at Rand leases (V) G.M. Co. Ltd. *Journal of the Chemical, Mining and Metallurgical Society of South Africa* 53:261–278.

Denver Sala Basic. 1994. *Selection Guide for Process Equipment*. Colorado Springs, CO: Denver Equipment.

Dickinson, H.W. 1945. Development of the stonebreaker. *Cement, Lime and Gravel* September:78–83.

Dinis Da Gama, C. 1983. Use of comminution theory to predict fragmentation of jointed rock masses subjected to blasting. *Transactions of the 1st Symposium on Rock Fragmentation by Blasting* 565–579.

The Domesday Book Online. 1999–2004. http://www.domesdaybook.co.uk/. Accessed July 2004.

Dorr, G.H. 1941. J.V.N. Dorr, the man. Address at Chemists' Club NYC, Award of Society of Chemical Industry's Perkin Medal to J.V.N. Dorr. January 10.

Dorr, J.V.N. 1936. *Cyanidation and Concentration of Gold and Silver Ores*. New York: McGraw-Hill.

Doughty, F.T.C. 1948. The cyclone–its use for mineral concentration. *Mine & Quarry Engineering* 14:337–340.

Drais Mannheim. 2000. *DCP Superflow*. Company brochure. Mannheim, Germany: Draiswerke.

Drier, R.W., and O.J. Du Temple. 1961. Prehistoric Copper mining in the Lake Superior region. A collection of reference articles. Michigan: Calumet.

Driessen, M.G. 1948. The use of cyclones as thickeners and washers in modern coal preparation. *Transactions of AIME* 177:240.

Drinker, H.S. 1888. *Tunneling, Explosive Compounds, and Rock Drills*. 3rd edition. New York: John Wiley & Sons.

Duda, W.H. 1987. *Cement Data Handbook*. Berlin: Bauverlag Weisbaden.

DuPont. 1980. *Blasters' Handbook*. 16th edition. Wilmington, DE: Explosive Products Division, EI du Pont de Nemours & Co. Inc.

Dyer, F.L., and T.C. Martin. 1910. *Edison: His Life and Inventions*. New York: Harper and Brothers Publishers.

Eckel, E.C. 1905. *Cements, Limes and Plasters*. New York: John Wiley & Sons.

Eckmeki, Z. 2003. Professor, Hacettepe University, Ankara, Turkey. Personal communication.

Eissler, M. 1897. *A Handbook of Modern Explosives*. London: Crosby, Lockwood and Sons.

Emmett, R.C., and D.A. Dahlstrom. 1953. Application of centrifugal forces to gravitational classifiers. *Transactions of AIME* 196(5):1015.

Evans, O. 1785. *The Young Mill-Wright and Millers Guide*. http://www.uh.edu/engines/epi285.htm. Accessed January 2004.

——. 1805. *The Abortion of the Young Steam Engineer's Guide*. http://www.history.rochester.edu/steam/evans/1805/. Accessed December 2003.

Fairless, M. c1900. *The Roadmender*. London and Glasgow: Collins Clear-Type Press.

Favreau, R.F. 1970. Blasting theories. *Proceedings of the Commercial Explosives Symposium*. Ste. Adele, Quebec: Canadian Industries Ltd.

Feige, F. 1993. The roller grinding mill—current technical position and potential for development. *Zement Kalk Gips* 8:E287–293.

Finster, H. 2002. Rhine area. Lignite industry in the area between Aachen and Cologne. http://www.hfinster.de. Accessed 2002.

Fisch, H. 2000. Experience with an innovative grinding concept at Morning Star Cement Hon Chong, Vietnam. *Proceedings Loesche Symposium*. Dusseldorf (June). Dusseldorf, Germany: Loesche GmbH.

Fischer, E.K. 1944. The evolution of mills for grinding. *Interchemical Review* Winter:91–104.

Fisher, G.R. 1937. *Staff Reports* on Overseas Trips, 1937. Brisbane, Australia: University of Queensland Library.

Fitch, E.B., and E.C. Johnson. 1953. Operating behaviour of liquid-solid cyclones. *Transactions of AIME* 196(5):304.

Fontein, F.J. 1961. *Separation by Cyclone According to Specific Gravity Cyclones in Industry*. London: Elsevier.

Forbes, R.J. 1955. *Studies in Ancient Technology*. Volume 3. Lieden, The Netherlands: E.J. Brill.

Forster, A., and J. Reilly. 1922. The colloid mill. *Journal Society Chemical Industry, London* 41(20):435–438.

Foster, C.L.N. 1894. *A Textbook of Ore and Stone Mining*. London: Charles Griffin.

Fuller Company. 2001. *O-Sepa High Performance Air Separator*. Company brochure. Bethlehem, PA: Fuller Company, a member of the F.L. Smidth-Fuller Engineering Group.

Gardco. 2004. *Dispersion: Fineness of Grind Gages*. Pompano Beach, FL: Paul N. Gardner Co.

Gelberg, K.U., and K. Lange. 1998. *1898–1998, the History of Alpine*. Augsburg, Germany: Hosokawa Alpine Aktiengesellschaft.

Gille, B. 1966. *The Renaissance Engineers*. London: Lund Hopkins.

Gimpel, J. 1976. *The Medieval Machine*. New York: Holt, Rinehart and Winston.

Gipe, P. 1995. *Wind Engineering Comes of Age*. New York: John Wiley & Sons.

Godn, B., and C. Wilm, editors. 1994. *Primary Cereal Processing: A Comprehensive Sourcebook*. New York: VCH.

Good News Bible. 1976. London: Collins/Fontana.

Goren-Inbar, N., G. Sharon, M. Kislev, and Y. Melamed. 2002. Remains of seven types of edible nuts and nutcrackers found at 780,000-year-old site. Reported by H.J. Gleit. http://www.eurekalert.org/pub_releases/2002-02/huoj-ros021702.php. Accessed February 2003.

Grimm Brothers. 1993. The juniper tree. In *Grimm's Fairy Tales.* London: Tiger Books International.

Grimshaw, V.C., and J.F. Albus. 1983. Fluid energy grinding and drying for fine powder production. In *Ultrafine Grinding and Separation of Industrial Minerals.* New York: SME-AIME.

Halasz, N. 1959. *Nobel: A Biography.* London: Robert Hale Ltd.

Hall, R.G. 1935. Hadsel mill. *Transactions of AIME* 112:15–24.

Hardgrove, R.M. 1938. Effects of grindability on particle size distribution. *Transactions of AIChE* 34:131–152.

Hardinge, H.W. 1955. Making rock grind itself. *Engineering and Mining Journal* 156:84–90.

Harper's New Monthly Magazine. 1871. The Mount Cenis railway and tunnel. XLIII(CCLIV) (July).

Harries, G. 1977. The calculation of the fragmentation of rock from cratering. *Proceedings of the 15th International APCOM Symposium.* Melbourne: Australasian Institute of Mining and Metallurgy.

Hartman, G. 1940. *The World We Live In.* Melbourne: McMillan Co.

Healy, J.F. 1978. *Mining and Metallurgy in the Greek and Roman World.* London: Thames and Hudson.

Heiskanen, K. 1993. *Particle Classification*. London: Chapman and Hall.

Hicom International. 2001. *High Intensity Grinding Mills.* Company brochure. Sydney: Hicom International.

Hochscheid, R.E. 1955. Difficult classification problems solved with the DSM cyclone. *Journal of the Chemical, Mining and Metallurgical Society of South Africa* 56:197.

Hodges, H. 1970. *Technology in the Ancient World*. London: Penguin Press.

Hopkins, K. 1978. *Conquerors and Slaves*. Cambridge, UK: Cambridge University Press.

Hosokawa Alpine. 1999. *The New Generation Fluidised Bed Jet Mills.* Company brochure. Augsburg, Germany: Hosokawa Alpine Aktiengesellschaft.

Hukki, R.T. 1961. Proposal for a Solomonic settlement between the theories of von Rittinger, Kick and Bond. *Transactions of SME-AIME* 220:403–408.

Infoplease.com. 2003. Total population of the world by decade, 1950–2050. http://www.infoplease.com/ipa/A0762181.html. Accessed December 2003.

Ingersoll-Rand Company. 1995. *The Hole Story.* Company brochure. Montvale, NJ: Ingersoll-Rand.

Jacobs, H.E. 1944. *6000 Years of Bread.* New York: Doubleday Dorn and Co.

JKMRC. 2001. *Hybrid Stress Blasting Model.* A De Beers feasibility study. Brisbane, Australia: JKMRC.

Jolliffe, J. 2001. St. Piran's Day–March 5. http://www.celticwomen.org/article7.html. Accessed January 2004.

Karrer, P. 1947. *Organic Chemistry.* Translated by A.J. Mee. Amsterdam: Elsevier.

Kelsall, D.F. 1952. A study of the motion of solid particles in a hydraulic cyclone. *Transactions of the Institution of Chemical Engineers* 30:87.

——. 1953. A further study of the hydraulic cyclone. *Chemical Engineering Science* 2:254–272.

Khodorovskiy, M. 2000. Biography of Melvin A. Cook. http://www.slcc.edu/schools/hum_sci/physics/whatis/biography/cook.html. Accessed May 2003.

Kick, F. 1883. The law of proportional resistance and its application to sand and explosions. *Dinglers Journal* 250:141–145.

Kleine, T.H. 1988. *A mathematical model of rock breakage by blasting.* Ph.D. dissertation. University of Queensland, Australia.

Klumpar, I.V., F.N. Currier, and T.A. Ring. 1986. Air classifiers. *Chemical Engineering* March (3):77–82.

Köppern Equipment Inc. 2002. High pressure grinding with roller presses. http://www.koppernusa.com/grinding.asp. Accessed January 2004.

Kozmin, P.A. 1917. *Flour Milling.* London: George Routledge & Sons.

K.R. Komarek Inc. 2002. The briquetting process. http://www.komarek.com. Accessed December 2003.

Kubota Tower Mill Corp. 1998. *Latest Type of Fine and Ultrafine Grinding with Tower Mills.* Company brochure. Tokyo: Kubota.

Kuprits, Y.N., editor. 1967. *Technology of Grain Processing and Provender Milling.* Jerusalem: Israel Program for Scientific Translations.

Kurtz, B.E., and A.J. Barduhn. 1960. Compacting granular solids. *Chemical Engineering Progress* 56(1).

Kuznetsov, V.M. 1973. The mean diameter of fragments formed by blasting rock. *Soviet Mineral Science* 9(2):144–148.

Law, T.M., J. May, A.T. Spathis, A.T. Du Plessis, and A.M. Palmer. 2001. Blast damage and blast dilution control: The application of bulk emulsion systems at WMC St. Ives Junction Mine. *Fragblast* 5(1–2):1–20.

Lebeter, F. 1949a. Historical survey of crushing and grinding. *Mine & Quarry Engineering* September.

——. 1949b. The development of crushing and grinding. *Mine & Quarry Engineering* December:385.

Lee, H.-W., and A. Süssegger. 1995. Two years of experience with the V-separator. *World Cement* November:13.

Leonard, J.W., III. 1991. *Coal Preparation.* Littleton, CO: SME.

Lewis, F.H. 1898. *Engineering Record* 39:47–49.

Lienhard, J.H. 1988–1997. The engines of our ingenuity. No. 285, Oliver Evans. http://www.uh.edu/engines/epi285.htm. Accessed January 2004.

Lloyd, P. 2001. Consultant. Personal communication.

Lloyd, P.J.D., A.A. Bradley, A.L. Hinde, K.H. Stanton, and K. Schymura. 1982. Centrifugal milling on a commercial scale. *Engineering and Mining Journal* December:49–54.

Lock, C.G.W. 1889. *Practical Gold Mining.* London: E and FN Spon.

Lougheed, A.L. 1985. *The Cyanide Process and Gold Extraction in Australia 1888–1913.* Brisbane, Australia: University of Queensland, Fryer Library.

Louis, H. 1894. *A Handbook of Gold Milling.* London: McMillan and Company.

——. 1909. *The Dressing of Minerals.* London: Edward Arnold.

Lynch, M.L. 2002. *Mining in World History.* London: Reaktion Books Ltd.

Magner, L.N. 1992. *A History of Medicine.* New York: Marcel Dekker.

McAdam, R., and R. Westwater. 1958. *Mining Explosives.* London: Oliver & Boyd.

McAndrew, B. 2000. Oliver Evans. http://www.dickinson.edu/carlisle/barnitz/mcandrew.html. Accessed February 2004.

McCaffrey, D.J.A. 2003. Technical Officer, British Coal Utilisation Research Association. Personal communication. See also http://www.bcura.org/intro.html. Accessed November 2003.

McGregor, K. 1967. *The Drilling of Rock*. London: CR Books.

McGrew, B. 1950. Crushing practice and theory. First published in *Rock Products*. Reprint, New York: Maclean-Hunter Publishing Company.

McQuiston, F.W., and R.S. Shoemaker. 1978. *Primary Crushing Plant Design*. New York: SME-AIME.

MIM Process Technologies. 2002. *An Introduction to Ultra Fine Grinding Technology—The IsaMill*. Company brochure. Brisbane, Australia: MIM Process Technologies.

Mosse, C. 1969. *The Ancient World at Work*. London: Chatto and Windus.

Napier-Munn, T.J., S. Morrell, R.D. Morrison, and T. Kojovic. 1996. *Mineral Comminution Circuits*. Australia: JKMRC, University of Queensland.

Naske, C. 1911. *Zerkleinerungsvorrichtungen und Mahlanlagen*. Leipzig: Verl Spanner.

Naylor, T.R. 1958. The hydrocyclone in the refining of China clay. *Mine & Quarry Engineering* 24.

Nef, J.U. 1958. *Cultural Foundations of Industrial Civilisation*. Cambridge, UK: Cambridge University Press.

Netzsch, Inc. 1974. *The Bead Mill: Its History and Principle of Operation*. Lionville, PA: Netzsch.

Neuberger, A. 1930. *The Technical Arts and Sciences of the Ancients*. London: Methuen.

Nitta, S. 1998. Manager, Kubota Tower Mill Corp., about the origin of the tower mill and the development process. Personal communication.

Nobel Foundation. 2004. http://www.nobel.se/nobel/nobel-foundation/. Accessed July 2004.

O'Brien, R. 1965. Machines. In *Life Science Library*. Amsterdam: Time-Life International (Nederland) NV.

Oner, M. 2002. Professor, Hacettepe University, Ankara, Turkey. Personal communication.

Ongtkoon, Y. 2000. Experience in adopting the largest vertical roller mill in the world. *Proceedings Loesche Symposium*. Dusseldorf: Loesche GmbH.

Onuma, E., and T. Furukawa. 1984. O'Sepa—A new high performance air classifier. *World Cement* January/February 15:13–24.

Orth, R.A., and J.A. Shellenberger. 1988. Origin, production and utilization of wheat. In *Wheat: Chemistry and Technology*. Volume 1. Edited by Y. Pomeranz. St. Paul, MN: American Association of Cereal Chemists.

Paul, W. 1970. *Mining Lore*. Portland, OR: Morris Print Co.

Peele, R., editor. 1941. *Mining Engineers Handbook*. Volume 1, Section 4. New York: John Wiley & Sons.

Perrogon, J.M. 1994. History and ethnology of primary processing. In *Primary Cereal Processing: A Comprehensive Sourcebook*. Edited by B. Godn and C. Wilm. New York: VCH.

Peterson, W.F. 1976. *An Industrial Heritage*. Milwaukee, WI: Milwaukee County Historical Society.

Planet Wheat—Kansas Wheat Commission. 2002a. A short history of bread. http://www.cyberspaceag.com/breadhistory.html. Accessed March 2002.

——. 2002b. A short history of flour milling. http://www.cyberspaceag.com/flourmillinghistory.html. Accessed March 2002.

Pollio, M.V. [25 BC]. 1931–1934. *On Architecture*. 2 volumes. Reprint. New York: G.P. Putnam's Sons.

Preece, D.S., R.P. Jensen, and S.H. Chung. 2001. Development and application of a 3-D rock blast computer modeling capability using discrete elements—DMCBLAST_3D. *Proceedings of the 27th Annual Conference of the International Society of Explosives Engineers* 1:11–18. Cleveland, OH: International Society of Explosives Engineers.

Probert, A. 1969. Bartolome de Medina: The patio process and the sixteenth century silver crisis. *Journal of the West* 8(1):90–124.

Ramelli, A. 1588. Le diverse et artificiose machine. Paris: Published by author.

Reschner, K. 2002. Scrap tire recycling. http://home.snafu.de/kurtr/str/en_sr.html. Accessed January 2004.

Research Association of the British Paint, Colour and Varnish Manufacturers. 1953. Corn and paint. Monthly memorandum 112:4.

Reynolds, J. 1970. *Windmills and Watermills*. New York: Praeger Publishers.

Reynolds, T.S. 1983. *Stronger than 100 Men: History of the Vertical Water Wheel*. Baltimore: John Hopkins University Press.

Richards, R.H. 1908. *Textbook of Ore Dressing*. 2nd edition, Volume 1. New York: McGraw-Hill.

Richards, R.H., and C.E. Locke. 1940. *Textbook of Ore Dressing*. 3rd edition. New York: McGraw-Hill.

Rickard, T.A. 1897. *Stamp Milling of Gold Ores*. London: Scientific Publishing Co.

Robinson, B. 1980. The history of autogenous grinding. *Bulletin of the Canadian Institute of Mining and Metallurgy* October:114–117.

Rodengen, J.L. 1995. *The Legend of Ingersoll Rand*. Fort Lauderdale, FL: Write Stuff Syndicate, Inc.

Roe, L.A. 2000. Letter to editor. *Mining Engineering* July:17.

Rose, T.K. 1906. *The Metallurgy of Gold*. London: Charles Griffin.

Rosin, P., and E. Rammler. 1933. The laws governing the fineness of powdered coal. *Journal of the Institute of Fuel* 7:29–36.

Rowland, C.A., and R.C. Nealy. 1969. Experts look at dry grinding rod mills. *Transactions of SME-AIME* June:85–90.

Rumpf, H. 1977. Problems of scientific development in particle technology, looked upon from a practical point of view. *Powder Technology* 18:3–17.

Schaefer, H.-U. 1999. Vertical roller mills for cement production. Presented at the International Symposium for Utilization of Metallurgical Slags (ISUS99), Beijing, November 16–19.

Scheibe, W. 1993. 100 years of tube ball mills—A historical review. *Zement Kalk Gips* 5:E133–135.

Schmeisser, K. 1898. *Goldfields of Australasia*. London: MacMillan and Company.

Schonert, K. 1967. Modellrechnungen für Zerkleinerungsprozesse mit den Ergebnissen der Einzelkorn-Druckzerkleinerung, 2. Europäisches Symposium Zerkleinern, Hrsg. H. Rumpf und W. Pietsch, DECHEMA Monographie Nr. 993–1026, S.241/280.

——. 1982. Method of fine and very fine comminution of materials having brittle behaviour. U.S. Patent 4,357,287. November 2, 1982.

——. 1988. A first survey of grinding with high compression roller mills. *International Journal of Mineral Processing* 22:401–412.

Schonert, K., K. Ohe, and H. Rumpf. 1965. Technische Feinzerkleinerung mit Einzelkornbeanspruchung zwischen zwei Flächen. *Chemie-Ingenieur-Technik* 37, S.259/264.

Schotz, S.P. 1931. Developing the colloid mill. *Crush and Grind* 1 (July–August):31.

Schubert, H., and E. Wachtler. 1987. Erich Rammler—A pioneer of particle technology. *Particle Characterisation* 4:45–48.

Scott, A., G. Chitombo, and T. Kleine. 1993. The challenge of the prediction and control of fragmentation in mining. *Proceedings of the Conference on Rock Fragmentation by Blasting*. Rotterdam: A.A. Balkema.

Seymour, H. 1924. *Crushing and Grinding Machinery*. London: Benn.

SME-AIME. 1985. *SME Mineral Processing Handbook*. New York: SME-AIME.

Smidth, F.L. 2000. http://www.flsmidth.dk/corporate/profile.htm. Accessed February 2001.

Steiner, H.J. 2000. The contribution of P. Rittinger to the science of mineral processing. *Proceedings of the 5th International Mining History Congress.* Milos Island, Greece: Milos Conference Center–George Eliopoulos.

——. 2002. Professor, University of Leoben, Austria. Personal communication.

Sticht, R.C. 1905. The development of pyritic smelting. *Proceedings of the Australasian Institute of Mining and Metallurgy* 11:1–70.

Stone, D.I., and S. MacKinnon. 1976. *Life on the Australian Goldfields.* Carlton, Victoria: Methuen of Australia.

Storck, J., and W.D. Teague. 1952. *Flour for Man's Bread.* Minneapolis, MN: University of Minnesota Press.

Sumberg, J. 1999. The Hoosac Tunnel on the Boston and Maine RR. http://www.intac.com/~jsumberg/hoosac.htm. Accessed April 2003.

Süssegger, A., and S. Strasser. 1996. Closed circuit grinding system. U.S. Patent 5,505,389. April 9.

Taggart, A.F. 1927. *Handbook of Ore Dressing.* New York: John Wiley & Sons.

Tarjan, G. 1981. *Mineral Processing*. Budapest: Akademiai Kiado.

Tepe, J.B., and W.K. Woods. 1943. *Design of Ether-Water Contacting System*. Washington, DC: Atomic Energy Commission.

Tiggesbaumker, P. 2001. 175 years of cement manufacturing history. *Proceedings of Krupp Polysius Autumn Seminar.* Beckum, Germany: Krupp Polysius.

Toussaint-Samat, M. 1994. *A History of Food.* Oxford, UK: Blackwell.

Travis, P.M. 1929. Mechanical dispersion by means of the colloid mill. *Industrial and Engineering Chemistry* 21(5):422.

Trawinski, H. 1955. Multi-hydrocyclones of the latest design. *Chemie Ingenieur Technik* 27:193.

Trivedi, B.P. 2002. Chimp nut-cracking site offers clues to early tool use. http://nationalgeographic.com/news/2002/05/0523_020523_0523TVchimps.html. Accessed February 2003.

Truscott, S.J. 1923. *A Textbook of Ore Dressing*. London: MacMillan and Co.

von Rittinger, P.R. 1867. *Textbook of Mineral Dressing (Lehrbuch der Aufbereitungskunde).* Berlin: Ernst and Korn.

Vozar, J., and H. Kunnert. 1972. Peter Ritter von Rittinger–His CV and career. *Proceedings of the Rittinger Symposium.* New York: Springer-Verlag.

Walton, J., editor. 1978. *The Josephine Mill and Its Owners.* Cape Town: Collectors Press.

Ward-Perkins, J., and A. Claridge. 1980. *Pompeii–AD. 79.* Sydney: Australian Gallery Directors Council Ltd.

West, P.C.H. 1910. *The Modern Manufacture of Portland Cement.* London: Crosby, Lockwood and Sons.

Wilkinson, N.B. 1966. Explosives in History: The Story of Black Powder. Wilmington, DE: Hagley Museum.

Wilson, E.B. 1908. *Cyanide Processes.* New York: John Wiley & Sons.

Wright, O.H. 1949. *Cyclones for Desliming and Dewatering in the Sand Pebble Phosphate Field.* Florida Section: AIME.

Wuth, D.E., and R.L. O'Shields. 1955. New mud de-sander cuts drilling costs. *Oil & Gas Journal* 54:87.

Valery, W. 2003. Manager, Metso Minerals, Brisbane, Australia. Personal communication.

Yildiz, N. 1999. *Grinding–Theory and Application of Mills and Classifiers.* Sirketi, Turkey: Kozan Ofset Matbaacilik ve Ticaret Ltd.

ZKG International. 2000. *Process Know How* 53(4).

Index

Note: *f.* indicates figure; *t.* indicates table.

About the Authors

Both authors have enjoyed long and distinguished careers in mining science and engineering. Alban Lynch's primary expertise is in the simulation and automatic control of mineral treatment processes, including the development of engineering models of communition and classification processes and the application of these models to the optimization and design of plant circuits. He began his career in the 1940s as an industrial chemist in Sydney, Australia, and later moved into metallurgy with The Zinc Corporation Ltd. in Broken Hill. In the late 1950s, he became affiliated with the Department of Mining and Metallurgical Engineering at the University of Queensland, where he held progressively responsible positions–culminating with department head–until 1993. From there, he went on to serve as a visiting professor at the Federal University of Minas Gerais in Belo Horizonte, Brazil; the University Sains Malaysia in Ipoh; the Universidad Autonoma de San Luis Potosi in Mexico; and the Hacettepe University in Ankara, Turkey. He holds DSc, MSc, and BSc degrees from the University of New South Wales, and a PhD from the University of Queensland. Lynch received numerous professional awards during his career, has lectured all around the world, and has published more than 200 technical papers.

Alban J. Lynch

Chester Rowland graduated from the New Mexico College of Agriculture and Mechanical Arts in Las Cruces (later renamed New Mexico State University) with a BS degree in mechanical engineering in 1945. He went to work for the Allis-Chalmers Manufacturing Company in Milwaukee, Wisconsin, where he had the opportunity to work with Fred Bond, a pioneer in the mining industry. Rowland worked in various positions with Allis-Chalmers, including supervisor of grinding mill applications and manager of grinding mill sales. When Fred Bond retired in 1964, Rowland took his place, becoming the company's consultant on grinding. Ultimately he was also manager of grinding mill applications and comminution research and development, a position that Rowland held until he retired in 1982. Since then, he has continued to work as an independent consultant on grinding mill selection and grinding process challenges. He is a distinguished member of SME, has received a number of prestigious awards for his outstanding contributions to the field of mineral science, and has published more than 50 professional papers. Rowland has visited mines throughout the world and presented technical lectures and seminars in the United States and abroad.

Chester A. Rowland